Handbook of Telecommunications

Handbook of Telecommunications

Edited by **Claude McMillan**

WILLFORD PRESS

New York

Published by Willford Press,
118-35 Queens Blvd., Suite 400,
Forest Hills, NY 11375, USA
www.willfordpress.com

Handbook of Telecommunications
Edited by Claude McMillan

International Standard Book Number: 978-1-68285-150-0 (Hardback)

The publisher's policy is to use permanent paper from mills that operate a sustainable forestry policy. Furthermore, the publisher ensures that the text paper and cover boards used have met acceptable environmental accreditation standards.

Trademark Notice: Registered trademark of products or corporate names are used only for explanation and identification without intent to infringe.

Printed in the United States of America.

Contents

Preface

This book was inspired by the evolution of our times; to answer the curiosity of inquisitive minds. Many developments have occurred across the globe in the recent past which has transformed the progress in the field.

Telecommunication is the most widely used platform for exchange of information via channels such as electromagnetic waves, etc. This book on telecommunications seeks to educate the readers on significant topics such as WiMAX technology, mobile and wireless computing, algorithms and modeling, etc. It is a complete source of knowledge on the present status of this important field. With state-of-the-art inputs by acclaimed experts, this book targets students and professionals alike.

This book was developed from a mere concept to drafts to chapters and finally compiled together as a complete text to benefit the readers across all nations. To ensure the quality of the content we instilled two significant steps in our procedure. The first was to appoint an editorial team that would verify the data and statistics provided in the book and also select the most appropriate and valuable contributions from the plentiful contributions we received from authors worldwide. The next step was to appoint an expert of the topic as the Editor-in-Chief, who would head the project and finally make the necessary amendments and modifications to make the text reader-friendly. I was then commissioned to examine all the material to present the topics in the most comprehensible and productive format.

I would like to take this opportunity to thank all the contributing authors who were supportive enough to contribute their time and knowledge to this project. I also wish to convey my regards to my family who have been extremely supportive during the entire project.

Editor

PERFORMANCE EVALUATION OF REALISTIC VANET USING TRAFFIC LIGHT SCENARIO

Nidhi[1] and D.K. Lobiyal[2]

[1,2]School of Computer and Systems Sciences, Jawaharlal Nehru University, New Delhi, India
nidhi.jjmsn@gmail.com, lobiyal@gmail.com

ABSTRACT

Vehicular Ad-hoc Networks (VANETs) is attracting considerable attention from the research community and the automotive industry to improve the services of Intelligent Transportation System (ITS). As today's transportation system faces serious challenges in terms of road safety, efficiency, and environmental friendliness, the idea of so called "ITS" has emerged. Due to the expensive cost of deployment and complexity of implementing such a system in real world, research in VANET relies on simulation. This paper attempts to evaluate the performance of VANET in a realistic environment. The paper contributes by generating a real world road Map of JNU using existing Google Earth and GIS tools. Traffic data from a limited region of road Map is collected to capture the realistic mobility. In this work, the entire region has been divided into various smaller routes. The realistic mobility model used here considers the driver's route choice at the run time. It also studies the clustering effect caused by traffic lights used at the intersection to regulate traffic movement at different directions. Finally, the performance of the VANET is evaluated in terms of average delivery ratio, packet loss, and router drop as statistical measures for driver route choice with traffic light scenario. This experiment has provided insight into the performance of vehicular traffic communication for a small realistic scenario.

KEYWORDS

Intelligent Transportation System, Vehicular Ad-hoc Networks, Geographical Information System, Mobility Model Generator for Vehicular Networks[v.2.9], Simulation of Urban Mobility[0.12.3], Network Simulator-2.34.

1. INTRODUCTION

As per the World Health Organization (WHO) statistics, more than 1.3 million people worldwide are estimated to be killed each year out of road accidents. According to an online article published in Deutsche Welle [2] by Murali Krishnan dated 29.04.2010, "India's record in deaths has touched a new low, as toll rose to at least 14 deaths per hour in 2009 against 13 the previous year". While trucks/lorries and two-wheelers were responsible for over 40% deaths, the rush during afternoon and evening hours were the most fatal phases.[2,3]. Also, as per another article of WHO (article in Times of India, Dipak Kumar Dash, TNN, Aug 17, 2009, 04.10am IST) India leads the world in road deaths. In addition to this, some of the common problems to tackle with are the "Miles of Traffic Jam" on highway and the "Search for best Parking Lot" in an unknown city. [1]

For all the above mentioned reasons, the Government and Automotive Industries today pay lot of attention towards traffic management and regulation of a smooth traffic. They are investing many resources to slow down the adverse effect of transportation on environment, thereby increasing traffic efficiency and road safety. The advancements in technology, in the areas of Information and Communications, have opened a new range of possibilities. One of the most promising areas is the study of the communication among vehicles and Road Side Units

(RSUs), which lead to the emergence of Vehicular Network or Vehicular Ad-hoc Network (VANET) into picture. [1,4].

VANET is characterized as a special class of Mobile Ad hoc Networks (MANETs) which consists of number of vehicles with the capability of communicating with each other without a fixed infrastructure. The goal of VANET research is to develop a vehicular communication system to enable 'quick' and 'cost-efficient' transmission of data for the benefit of passenger's safety and comfort. Due to the expensive cost of deploying and complexity of implementing such a system in real world, research in VANET relies on simulation. However, the simulation depends on the mobility model that represents the movement pattern of mobile users including its location, velocity and acceleration over time. A mobility model needs to be a Realistic Mobility Model that considers the characteristics of the real world scenario either by taking a real world MAP obtained from TIGER(Topologically Integrated Geographic Encoding and Referencing) database from U.S. Census Bureau or by taking Satellite images of Google Earth into consideration to simulate a realistic network.[1]

In V2V communication or Inter-Vehicle communication, Vehicles are able to communicate with other ongoing vehicles on their path . In this scenario, it is not known in advance when it is possible to meet another vehicle to which the communication is feasible. In V2I communication, vehicles are able to communicate with Road Side Unit (RSUs) or access points. Whereas, In Inter-Road Side Communication it is possible to know the communicating parties in advance as RSUs are placed at a fixed distance from each other. Therefore, the main difference between V2V and V2I is the coverage area[8,21].

Rest of the paper is organized as follows. Related work is briefly described in Section 2. In Section 3, the methodology of proposed work is explained along with various tools which are used to carry out the work. Section 4 further discusses the simulation of network, results and analysis obtained through simulations conducted. Finally, Section 5 concludes the work presented in this paper.

2. RELATED WORK

Research is being carried out in the field of VANET such as Analyzing data dissemination in VANETs, Identifying and studying routing protocols in VANET in terms of highest delivery ratio and lowest end-to-end delay etc. The issues of Security and Privacy also demands great attention. The study of Mobility Models and their realistic vehicular model deployment is a challenging task.[8] Random way Point(RWP)[9] is an earlier mobility model widely used in MANET in which nodes move freely in a predefined area but without considering any obstacle in that area. However, in a VANET environment vehicles are typically restricted by streets, traffic light and obstacles. Ana et. al. [20] considers the mobility model in which vehicles know from the start their initial and final points. The routing track is then chosen considering the social relation between the vehicles and also the destination point. This means that vehicles move only between the initial and final point on path chosen by the social relation strength between the vehicles. GrooveSim [10] was the first tool for forecasting vehicular traffic flow and evaluating vehicular performance. It gives a traffic simulator environment which is easy to use for generating real traffic scenario for evaluation. But it fails to include network simulator as it was unable to create traces for network. David R. Choffnes et al. [11] proposed a mobility model named STRAW (Street RAndom Waypoint). This model has taken real map data of US cities and considered the node (vehicle) movement on streets based on this map. This model also has the functionality to simplify the traffic congestion by controlling the vehicular mobility. But still it lacks overtaking criteria that cause convey effect in street as it considered random method which is not realistic. Kun-chan Lan et al.[6] describes a realistic tool MOVE for generating realistic vehicular mobility model. It is built on top of an open source micro-traffic simulator SUMO and its output is a realistic mobility model that can immediately be used by popular network simulators such as ns-2 and qualnet. In paper [1] Kun-chan's MOVE model (v.2.81) was used to obtain the results of driver's route choice at the intersection without considering the traffic light scenario.

3. PROPOSED WORK & METHODOLOGY

To evaluate the performance of VANET, there is a need to deploy a real world scenario with all the vehicular constraints. In this paper the experiment was performed by taking a limited bounded region of a real world scenario i.e. "JAWAHARLAL NEHRU UNIVERSITY (JNU), NEW DELHI, INDIA" into consideration. The steps to implement a VANET simulation in this region are as follows:

- Generation of JNU Map
- Creation of Vehicular Traffic flow on this Map
- Simulation of Network with traffic lights at Intersections

The detailed procedure in implementing such above mentioned steps are explained in the rest of this paper.

3.1. JNU Map Generation

For creating a real world Map of JNU, Some of the existing tools have been used such as Google Earth, ArcGIS 9 (ArcMap version 9.1), MOVE Simulator (v 2.9)[5,6] and Adobe Dreamweaver CS4. Satellite image of JNU has been taken from Google Earth shown in Figure 1. This image was further imported into ArcGIS 9 as depicted in Figure 2. **ArcGIS** is basically a suite consisting of a group of Geographic Information System (GIS) software products.[12]

NOTE: Google Earth gives latitude and longitude of a particular location whereas ArcGIS maps those latitudes and longitudes to the required coordinate plane with the desired origin in a Two Dimensional Space.

Some of the 2-D Co-ordinates of this Map were not lying in the first quadrant of the 2-D Co-ordinate plane. In order to obtain all the co-ordinates in the first quadrant, the origin was shifted to an appropriate location. Shifting of the old Co-ordinates (x, y) to a new origin (h, k) is given by :

$$X= x + h; Y= y + k ;$$

Where (X,Y) represents the translated Co-ordinates in the plane with new origin and the traffic lights at particular co-ordinates of intersection have been used as the inputs to the Map Node Editor of MOVE Simulator as shown in Figure 3. After creating nodes and traffic lights at particular intersections using Map Node editor, numbers of parameters are defined such as edges between nodes, number of lanes, speed and priority of roads on which vehicle move, with the help of Road Editor of MOVE simulator as shown in Figure 4. Here, a multi-lane scenario of two lanes with 75% road priority has been set. The threshold speed has been considered for each lane in a region of JNU Map as 40m/s. Next a connection was established between nodes via edges by writing an XML code (**.con.xml**)[16] using Dreamweaver CS4. Finally the nodes, edges and connection files are configured into **.net.xml** by using NETCONVERT to create the MAP. Figure 5 depicts the JNU Map created by the above defined tools.

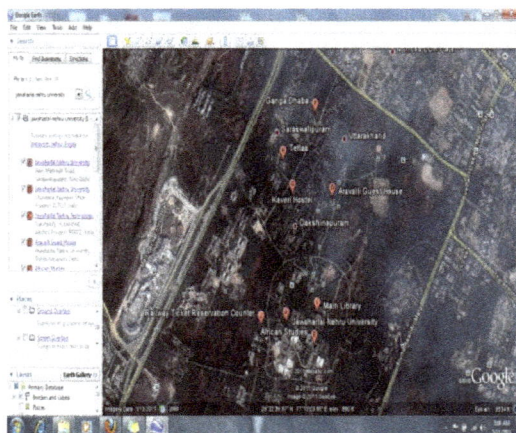

Figure 1 Satellite Image of JNU.

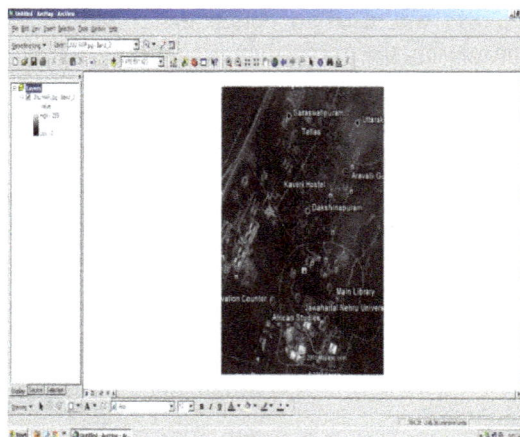

Figure 2 Imported Image of JNU in ArcGIS

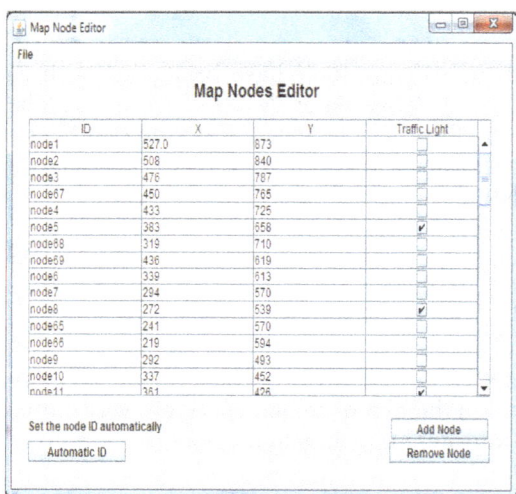

Figure 3. Map Node Editor of MOVE

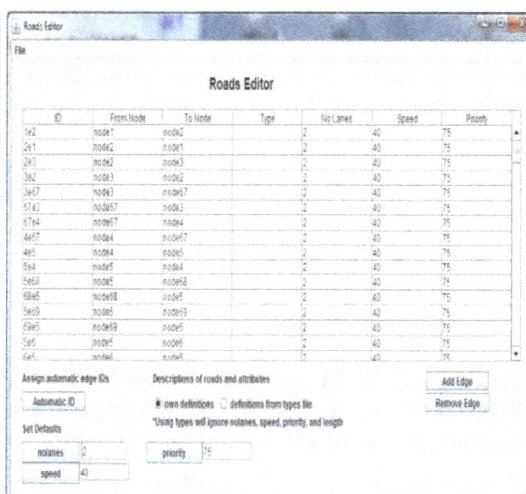

Figure 4. Road Editor of MOVE

3.2 Traffic Flow

For generating a vehicular traffic flow by taking traffic lights into consideration on the above created Map, SUMO 0.12.3[17] simulator has been used in addition to MOVE simulator. Initially, the Route File in XML (**rou.xml**) was created, in which acceleration, deceleration, maximum speed, length and type of a vehicle were specified (see Table 1). In addition to this, the bounded JNU region has been divided into 36 smaller routes which the vehicles can take. Further, the departure time of a particular vehicle on a particular route which creates the vehicular traffic flow among the nodes has been specified. The vehicle's destination from the source and their turning directions at the intersections, such as right turn, left turn and straight as per their destination were also set as per the driver's route choice at intersection. In addition to this, traffic light scenario has been taken into consideration at all the intersections to regulate the traffic movement in different direction as well as to analyze the clustering effect at the intersections.

Table 1 Types of Vehicle and their Characteristics

Vehicle Type	Max. Acc. (m/s^2)	Max. Dec. (m/s^2)	Length (m)	Max. Speed (m/s)
Car A	3.0	6.0	5.0	30
Car B	2.0	6.0	7.5	30
Car C	1.0	5.0	5.0	20
Car D	1.0	5.0	7.5	10

Different Route files with traffic lights at the intersections have been created for varying traffic flow consisting of 10,20,30,40,50,60,70 vehicles. This varying flow has been set by keeping in mind, a constant deceleration and acceleration model in which vehicles do not move and stop abruptly. Map file (.net.xml) and the different Route files (rou.xml) of varying traffic flow were configured to create the corresponding trace files (**sumo.tr**) which can be visualized using SUMO simulator. These trace files basically shows the JNU Map as shown in Figure 5 and the flow of traffic and clustering effects of vehicles due to traffic lights at the intersection are depicted in Figure 6(a) and (b). After setting the parameters of SUMO, the real world scenario of JNU region can be visualized with vehicles moving on it as depicted in Figure 6 (a), (b) & (c).

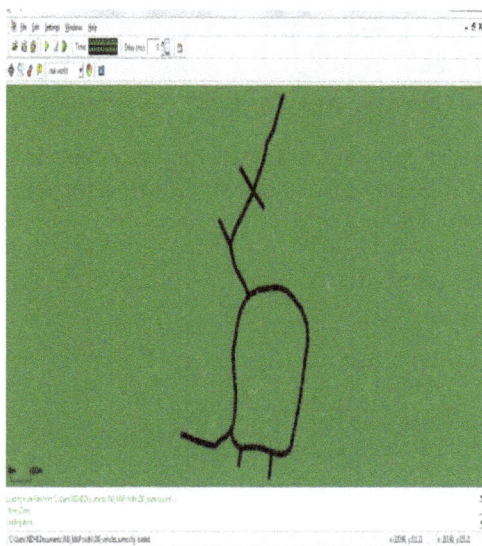

Figure 5. SUMO visualization of JNU Map .

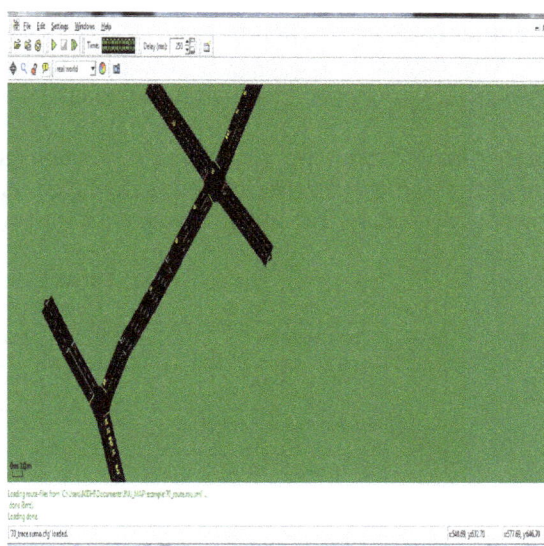

Figure 6(a). Vehicular Flow and Clustering of vehicles due to traffic light at the intersection

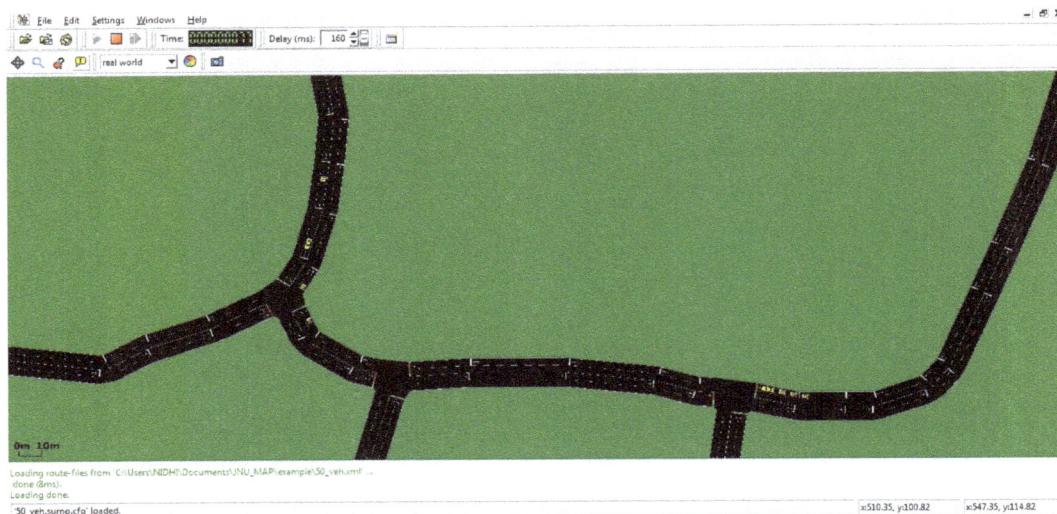

Figure 6(b)

4. TRAFFIC LIGHT SIMULATION

In order to simulate the clustering effects of vehicles, traffic lights have been installed in the network at the intersections shown in Figure 6(a) and (b). The traffic flow among the vehicles is generated using Traffic Model Generator of MOVE and Network Simulator (NS-2.34)[18].

The Traffic Model Generator of MOVE creates the dynamic mobility of varying number of vehicular traffic by generating traffic simulation file for simulation. The traffic simulation files have been generated by interfacing traffic flow with traffic lights created in section 3.2 with the JNU Map created in section 3.1. These traffic simulation files of different traffic scenario are subsequently used in NS-2 which facilitated the simulation of traffic flow of region under study.

Various parameters were considered for establishing the communication among vehicles. For example, a vehicular traffic flow was deployed using IEEE 802.11 standard with transmission range of 250 meters. The other parameters used are discussed in Table 2.

Table 2. Network parameters

Parameters	Values
Channel Type	Wireless Channel
Propagation Model	Two Ray Ground Model
Network Interface Type	Wireless Phy
MAC Type	802.11
Interface queue	DropTail/Pri Queue
Link Layer Type	LL
Anetnna	Omni Antenna
Ifqlen	50
Varying No. of Nodes	10,20,30,40,50,60,70

Routing Protocol	AODV
Topology (X,Y) Co-ordinates	(659, 911)
Transmit Power, Pt	0.2818
Channel Frequency	2412e+6
RXThresh	3.65262e-10
CSThresh	(Expr 0.9 * RXThresh)

The simulation covers 600349 m^2 area and the following parameters have been used for traffic flow between nodes in our simulations.

.

Table 3. Parameters of Traffic Flow between nodes

Parameters	Values
Agent	UDP
Packet_size	1000
Application_Traffic	CBR
CBR Rate	64kbps
CBR_max_pkts	2280000
CBR interval	0.05micro sec
Different RNG seed	2,4,6,8,10

After setting up the network and traffic flow as discussed above, simulation was conducted by taking 1/4th of the vehicular nodes sending CBR traffic for a traffic scenario of 10 vehicles initially. In this scenario, 50% of the vehicles are involved in direct communication, whereas, rest 50% vehicles serves as the intermediate (or router) nodes for the communication. Further, all the traffic parameters as given in table 3 were kept constant except number of nodes sending CBR traffic. This we have considered always as 1/4th of the number of vehicles 20,30,40,50, 60 and 70 respectively.

4.1 Simulation Results

The impact of realistic vehicular mobility (using various tools as discussed in Section 3), on the performance of ad-hoc routing protocols has been evaluated in this section.

The driver route choice behaviour with traffic lights at the intersections has been simulated for a real world scenario. In this, all possible routes from the source to destination are defined and the driver needs to decide about which route is to be taken from among all possible routes at any intersection. The presence of traffic lights at the intersection regulates the smooth movement of vehicles in different directions and causes clustering effect by forcing the vehicles to stop at intersection when the signal is red. Therefore, the node density at the intersection increased which improves the network connectivity among the peers at intersection, but the improved connectivity deteriorates the packet delivery ratio.

Our simulation concentrates on selecting the probability of choosing a route at the intersection. This probability directly determines the number of vehicles on a particular route. The data in terms of packets are transmitted to facilitate communication among vehicles. In order to study the behaviour of communication, the parameters like delivery ratio, packet loss and router drop has been considered which are discussed in the subsequent sections.

4.1.1 Average Delivery Ratio

Delivery Ratio implies the ratio of number of packets successfully delivered to the number of packets sent.
For calculating delivery ratio with respect to the number of vehicles, different traffic scenarios were simulated with varying number of vehicles in multiples of 10. For each scenario, delivery ratio was calculated for 5 simulation runs by changing the seed in multiples of 2. The Average delivery ratio for each scenario was an average
of 5 simulation runs and it is calculated as follows:

$$APR = (\sum_{k=1}^{5} PR) / 5$$

$$APS = (\sum_{k=1}^{5} PS) / 5$$

$$ADR \% = (APR/APS) * 100$$

Where, PR = Packet Received, PS = Packet Sent, APR = Average Packet Received and APS = Average Packet Sent.

The summary of results obtained is shown in table 4 and the results are further analyzed graphically in figure 7. It can be observed that the choice of route at intersection points can significantly affect the simulation results.

Table 4 Number of Traffic and Avg. Delivery Ratio (ADR) %

No. of Vehicular Traffic	Packet Delivery Ratio %
10	87.68 %
20	82.05 %
30	76.76 %
40	57.13 %
50	52.42 %
60	46.02 %
70	16.61 %

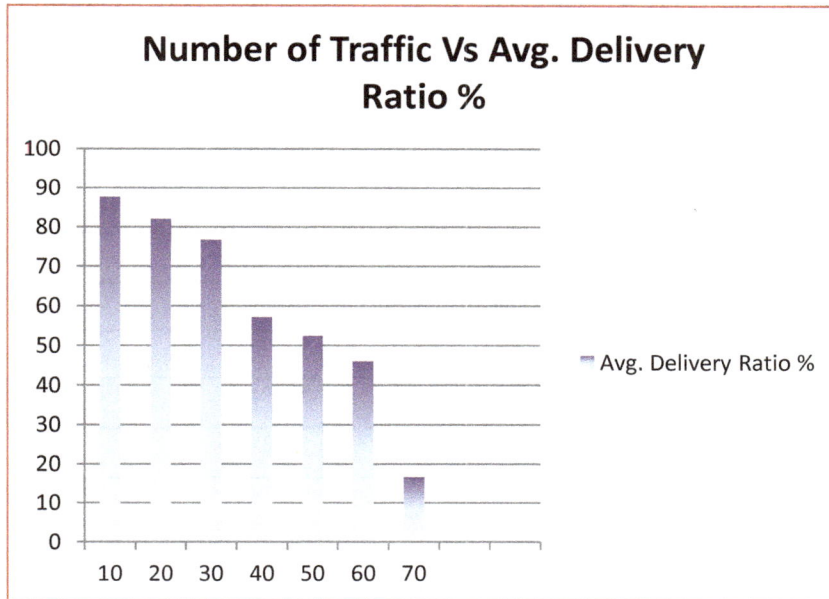

Figure 8. Number of Traffic Vs Avg. Delivery Ratio% using traffic light scenario.

It is clear from the figure 8 that the delivery ratio decreases as the node density increases due to the the increased collision caused by higher node density. However, as higher node density increases connectivity, but at the same time, it may result in more number of collisions.

4.1.2 Router Drop:

Router Drop for each traffic scenario is calculated by taking the average of Router Drop to the packets sent as shown below :

$$RD \% = \left(\sum_{K=1}^{5} \frac{RD}{PS} \right) * 100$$

Where, RD % = Router Drop %

4.1.3 Packet Loss:

Packet loss is calculated by taking the average of packet loss to the packets sent.

$$PL = (PS - PR)$$

$$PL \% = \left(\sum_{K=1}^{5} \frac{PL}{PS} \right) * 100$$

Where, PL = Packet Loss.

The results obtained for Router Drop and Packet Loss are summarized in table 5 for varying vehicular traffic. This is further illustrated in figure9.

Table 5 (Number of vehicular traffic) Vs (Router Drop and Packet Loss%)

No. of Vehicular Traffic	Router Drop %	Packet Loss %
10	12.31 %	12.32 %
20	06.71 %	17.95 %
30	23.09 %	23.23 %
40	36.97 %	42.86 %
50	40.62 %	47.58 %
60	41.28 %	53.98 %
70	55.45 %	83.39 %

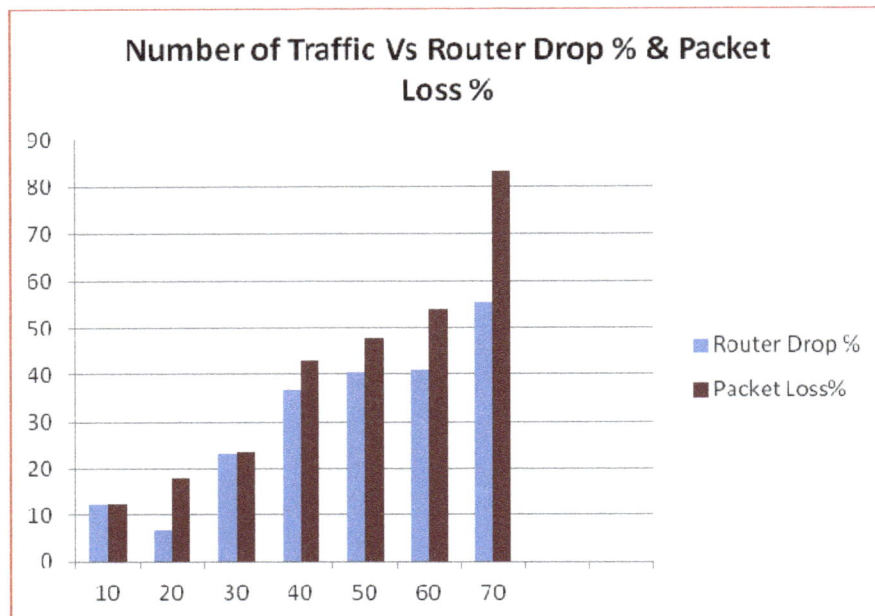

Figure 9. Vehicular Traffic Vs (Router Drop% & Packet Loss%) using traffic light scenario.

Our Simulation results suggest that, the presence of traffic lights at the intersections, deteriorate the packet transmission rate and increases the number of packet collision as shown in Figure 8 and 9. This happens due to the high node density at the intersection till the traffic light signal turns to green and it also results in collision of packets sent by vehicles at the same time. Further, it was observed that the packet delivery ratio kept on decreasing since the collision of packets completely depends on the clustering of vehicles at the intersection. This phenomenon can be explained by deployment and movement of vehicles in a given scenario. It seems that the connectivity between the vehicles get improved but simultaneously reduces the delivery ratio.

Figure 9 shows the effect of varying number of vehicles on packet loss and router drop together. A packet may be dropped by a vehicle or by a router. The packet loss percentage is considered as packets dropped by vehicles. Further in figure 9, it is quite evident that the percentage of packet loss was slightly more than router drops. This happens because of higher chances of packet being dropped at the end rather than being dropped at intermediate nodes. Here, again from the figure, it is quite evident that the router drop is not confined to any increasing or decreasing order. As explained for the case of lower deliver ratio in Figure 8, the drop rate was higher due to clustering effects of vehicles caused by traffic lights at the intersection.

5. CONCLUSION

In this paper, we have obtained an in-sight idea of simulating real world scenario of VANET. As it is not easy to deploy and implement such a complicated system in real world before knowing the impact of all parameters used in VANET, a small real world area i.e. our University, JNU itself, was taken into consideration, for studying the impact of mobility in the VANET. Traffic movement has been deployed across the area under consideration using one of the realistic vehicular mobility models. The behavior of this network was simulated using NS2 to study the impact of driver's choice with traffic lights at the intersection on packet transmission over V2V communication using AODV routing protocol and IEEE 802.11 standard.

The performance of the network has been evaluated by taking delivery ratio, packet loss and router drop as statistical measures. The average delivery ratio for various scenarios such as varying number of vehicles with constant power transmission range of 250m and frequency of 2.4GHz was observed to be 68.38% whereas packet loss was 40.18%.

Traffic light scenario has been an important measure to regulate the traffic flow in a round robin fashion. But for data transmission, it has become an obstacle since the packet forwarding nodes at the intersection drops the packets, due to the high number of transmission at the same time. Therefore, it is concluded from the results that packet drops may be reduced by using RSU's that can forward the packets even if the intermediate vehicles drops the packets at the intersections. Therefore, simulation of vehicular networks using RSU at the intersection will be our future direction of research.

REFERENCES

[1]   Nidhi & Lobiyal, D.K., (2012) "Performance Evaluation of VANET using realistic Vehicular Mobility", N. Meghanathan et al. (Eds.): Vol. 84, CCSIT , Part I, LNICST 84, pp. 477–489.

[2]   DW World-de: Deutsche Welle, http://www.dw-world.de/dw/article/0,,5519345,00.html

[3]   Khairnar, V.D., Pradhan, S.N,(2010) " Comparative Study of Simulation for Vehicular Ad-hoc Network" , IJCA (0975 – 8887) Volume 4– No.10.

[4]   Olariu, S., Weigh, M.C.,(2009) "Vehicular Networks, from theory to practice", Vehicular Network Book, CRC Press Publishers.

[5]   Huang, C.M., Chen, J.L., Chang, Y.C.,(2010) "Telematics Communication Technologies and Vehicular Networks", Wireless Architectures and Applications, Information Science Reference, New York Publisher.

[6]   Paier, A., Bernadó, L., Karedal, J., Klemp, O., Kwoczek, A.,(2010) "Overview of vehicle-to-vehicle radio channel measurements for collision avoidance applications", In: 71st IEEE Vehicular Technology Conference, VTC Spring Taipei.

[7]   Lan, K.C., Chou, C.M.(2008) "Realistic mobility models for Vehicular Ad hoc Network (VANET) simulations", In: 8th IEEE International Conference, pp. : 362 – 366,ITS Telecommunication.

[8]   Sichitiu M., Kihl M.,(2008) "Inter-vehicle communication systems: a survey", IEEE Communications Surveys & Tutorials, v. 10, n. pp. 88-105.

[9]   David R.C., Fabi´an E. B.(2005) "An integrated mobility and traffic model for vehicular wireless networks", In: 2nd ACM International Workshop on Vehicular Ad Hoc Networks (VANET), Cologne, Germany.

[10]   GIS Tutorial: How to Use ArcMap 9.1, http://www.trincoll.edu/depts/cc/documentation/GIS /HowToArcMap.pdf

[11]   Kim J., Sridhara V. , Bohacek S.(2008) "Realistic mobility simulation of urban mesh networks, Ad Hoc Networks", J. ScienceDirect, Elsevier, Ad Hoc Networks.

[12]   Hartenstein, H., Laberteaux, K.P.: VANET(2010) "Vehicular Applications and Inter-Networking Technologies", Vehicular Network Book, JohnWiley & Sons Ltd Publisher.

[13]   H¨arri J., Filali F., Bonnet C.(2007) "Mobility Models for Vehicular Ad Hoc Networks: A Survey and Taxonomy", Technical Report RR-06-168, Institut Eurecom .

[14]   Harold, E.R.(2001): XML Bible, XML Book,Hungry Minds Inc. 2nd edition.

[15]   SourceForge.net:sumo, Main Page,http://sourceforge.net /apps/mediawiki /sumo/ index.php? title=Main_Page

[16]   Pall, K., Vardhan K.: The ns Manual (formerly ns Notes and Documentation), http://www.isi.edu/nsnam/ns/doc/ns_doc.pdf

[17]   H¨arri J., Fiore M., Filali F., Bonnet C.,(2007) "A Realistic Mobility Simulator for Vehicular Ad Hoc Networks", Technical Report RR-05-150, Institut Eurecom.

[18]   XML Tutorial, http://www.w3schools.com/xml/

[19]   Preuss, M., Thomas, S.,(2008) "Wireless, mesh & ad hoc networks; Military convoy location and situation awareness", In: IEEE Sarnoff Symposium, pp. 1–5.

[20]   Gainaru, A., Dobre, C., Cristea, V.(2009) "A Realistic Mobility Model Based on Social Networks for the Simulation of VANETs", In: 69th IEEE Vehicular Technology Conference, pp. 1-5.

[21]   Kanitsorn Suriyapaiboonwattana, K., Pornavalai, C., & Chakraborty, G.,(2009) "An Adaptive Alert Message Dissemination Protocol for VANET to Improve Road Safety", FUZZ_IEEE.

[22]   Meyer, H., Cruces, O.T., Hess, A., Hummel, K., Ordinas, J.M.B., Casetti, C.E., Karlsson, G.(2011) "VANET Mobility Modeling Challenged by Feedback Loops", In: 10th Annual Mediterranean Ad Hoc Networking Workshop.

[23]   Sommer, C., Dietrich, I., Dressler, F.(2007) "Realistic Simulation of Network Protocols in VANET Scenarios", In: 26th IEEE INFOCOM, Mobile Networking for Vehicular Environments (MOVE), Poster Session.Anchorage, Alaska, USA

[24]   Yuwei, Xu., Ying, Wu., Gongyi, Wu., Jingdong, Xu., Boxing, L., Lin, S., (2010) "Data Collection for the Detection of Urban Traffic Congestion by VANETs", In: IEEE Services Computing Conference (APSCC), Volume: Issue: , 6-10 Dec., Asia-Pacific pp. 405 – 410.

[25]   Abdrabou, A., & Zhuang,W., "Probabilistic Delay Control and Road Side Unit Placement for Vehicular Ad Hoc Networks with Disrupted Connectivity", IEEE JOURNAL ON SELECTED AREAS IN COMMUNICATIONS,VOL. 29.

[26]   Djenouri D., Nekka E., Soualhi W.,(2008) "Simulation of Mobility Models in Vehicular Ad hoc Networks", 1st ICST On Ambient Media and Systems (Ambi-sys). Quebec, Canada.

[27]   Kone, V.:Data Dissemination in Vehicular Networks, www.cs.ucsb.edu/~vinod/docs /vinod _ mae.ppt

[28]   The Network Simulator- ns-2, http://isi.edu/nsnam/ns/

WiMAX Based 60 GHz Millimeter-Wave Communication for Intelligent Transport System Applications

[1]Bera Rabindranath, [2]Sarkar Subir Kumar, [3]Sharma Bikash, [4]Sur Samarendra Nath, [5]Bhaskar Debasish & [6]Bera Soumyasree

1,3,4,5,6 Sikkim Manipal Institute of Technology, Sikkim Manipal University, Majitar, Rangpo, East Sikkim, 737132:
2. Jadavpur University, Kolkata 700 032.

{rbera50,samar.sur,debasishbhaskar,soumyasree.bera}@gmail.com

Abstract

With the successful worldwide deployment of 3rd generation mobile communication, security aspects are ensured partly. Researchers are now looking for 4G mobile for its deployment with high data rate, enhanced security and reliability so that world should look for CALM, Continuous Air interface for Long and Medium range communication. This CALM will be a reliable high data rate secured mobile communication to be deployed for car to car communication (C2C) for safety application. This paper reviewed the WiMAX ,& 60 GHz RF carrier for C2C. The system is tested at SMIT laboratory with multimedia transmission and reception. With proper deployment of this 60 GHz system on vehicles, the existing commercial products for 802.11P will be required to be replaced or updated soon .

Key words: C2C, CALM , WiMAX, WiFi, VSG, RTSA .

1 Introduction

Safety and security are very important in car-to-car communication. It is even more important when wireless systems are used because it is generally perceived that wireless systems are easier to attack than wireline systems. In search of best, secured and reliable communication technology towards next generation e-car safety application, IEEE 802.16, an emerging wireless technology for deploying broadband wireless metropolitan area network (WMAN), is one the most promising wireless technology for the next-generation ubiquitous network. Though IEEE802.11P WiFi based products are commercially available for same functionality. But, disadvantages incurred in the Wi-Fi security have been addressed into the IEEE 802.16 standard and also flexibility parameters are also addressed in WiMAX. WiMAX is designed to deliver next-generation, high-speed mobile voice and data services and wireless "last-mile" backhaul connections [1]

The University of Texas at Austin. IEEE 802.16e (Mobile WiMax) deals with the Data Link Layer security. The Data-Link Layer authentication and authorization makes sure that the network is only accessed by permitted users while the encryption ensures privacy and protects traffic data from hacking or spying by unauthorized users. The WiMAX 802.16e provides number of advanced security protections including: strong traffic encryption, mutual device/user authentication, flexible key management protocol, control/ management message protection, and security protocol optimizations for fast handovers when users switch between different networks. Fig.1 shows a WiMAX architectural components

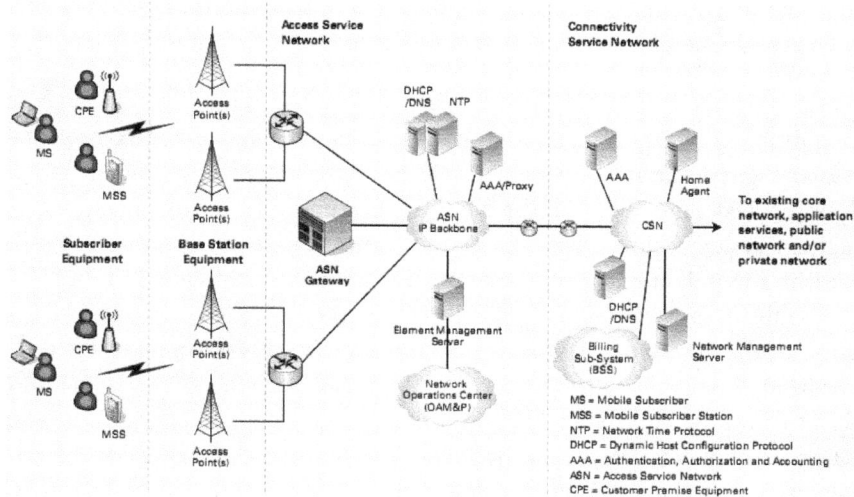

Fig. 1. WiMAX architectural components [2].

Commercial products of vehicular networks exists viz. DENSO's Wireless Safety Unit (WSU), Hitachi-Renesas.

DENSO's Wireless Safety Unit (WSU) is the follow up development to DENSO's first generation 802.11p communication module, the Wave Radio Module (WRM). It is specifically designed for automotive environments (temperatures, shock, vibration) and has its primary focus on safety related applications. [4]

During normal driving, the equipped vehicles anonymously share relevant information such as position, speed and heading. In a C2C environment message authorization is vital. The possibility to certify attributes and bind those to certain vehicles is particularly important for public safety. [5]

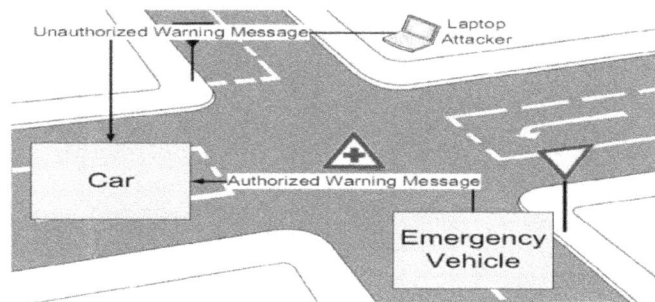

Fig. 2. Shows a possible attack in a typical Car2Car environment. Assuming no security, his attacker could generate valid messages for and consequently disturb the whole transportation system.

While unlicensed spectrum around 2.5 GHz and 5 GHz is also available internationally, the amount of available 60G bandwidth is much higher than that around 2.5GHz and 5GHz [3].

Unlicensed spectrum surrounding the 60 GHz carrier frequency has the ability to accommodate high-throughput wireless communications. It is highly directive and can be used for long and directed link. 60GHz system enjoys the size reduction and cost reduction advantages. Additionally, due to availability of 5GHz bandwidth the data-rate for communication is more interesting. Many commercial products have been developed facing these challenges.

Thus ,exploring the WiMAX 802.16e for its security, reliability and high throughput features , exploring 60 GHz millimeter wave as carrier for its size and cost reduction, wide bandwidth and highest throughput , the Car2Car communication system is required to be developed for the next generation Car for safety applications. The development of 60 GHz C2C communication system comprised of two step procedure discussed below. The MATLAB/SIMULINK is used for the design verification and simulation at the 1st stage. The final simulation result in the form of *.mdl file is ported to the ARB unit of one R&S VSG for the realization of the base band hardware. The transmit IF at 1 GHz and transmit RF at 60 GHz are realized through RF block of R&S VSG and separate 60 GHz transmit module respectively as shown in figure 8 and 9 respectively. The signal reception is developed using 60 GHz RF front end, RF Tuner and RTSA (Tektronix real Time spectrum analyzer). in Laptop, Data is retrieved at the Laptop using data acquisition and digital signal processing.

The above system development efforts are discussed below. Section 2 relates the mathematical modeling of sub-carrier generation. Section 3 will discuss all about WiMAX simulation at the base band level. The successful development of section 3 will produce one *.mdl file which is ported to the VSG for base band hardware realization.   Section 4 will discuss the efforts pertaining to hardware development.

2   Mathematical modeling of Sub-carrier generation

The serial data input is a sequence of samples occurring at interval T_s. At the transmitter as shown in Fig 3, the high rate serial input data is converted to column data which is of lower rate. The low-rate column data consists of M low-rate parallel streams in order to increase the symbol duration to $T=MT_s$. The low-rate streams, represented by the symbols $b_m[k]$, m=0, 1, 2, ..., M-1, and k=1, 2, 3......, where each stream is modulated onto different sub-carriers. The orthogonal relationship between any two of the sub-carriers in a set is maintained to avoid the Inter Channel Interference [8]. Then the parallel streams are multiplexed and a Cyclic Prefix is inserted to eliminate the effect of Inter Symbol Interference. So, we obtain the transmitted k^{th} symbol as below.

$$y(t) = \sum_{m=0}^{M-1} b_m[k]e^{j2\pi mt/T}, \text{ where, } -G+kT \le t \le (k+1)T \qquad (1)$$

G is the length of the Guard Interval and at the mth stream of data the kth symbol is coming out as $b_m[k]$.

Now this y(t) will pass through the channel which can be modeled as the frequency selective fading channel and it will also have the multipaths. This type of channel model can be realized if we consider the tapped-delay line with time-varying coefficients with a fixed tap spacing. The following is the mathematical realization of such type of channel [8].

$$h(t,\tau) = \sum_{l=0}^{\chi} h_l(t)\delta(\tau - \tau_l) \tag{2}$$

$h_l(t)$ and τ_l are the complex amplitude and delay of the lth path, respectively. $\chi+1$ is the total number of taps. τ_{χ} is the maximum multipath delay spread. For OFDM symbolization, the length of G should be greater than τ_{χ}. The expectation value of $h_l(t)$ is the determining factor to get the following correlation function. As $h_l(t)$ is random in nature, it is modeled as the Wide-Sense Stationary Uncorrelated Scattering process.

$$\phi_h(\Delta t) \overset{\Delta}{=} E[h_l(t)h_l^*(t - \Delta t)] = \sigma_l^2 \phi_t(\Delta t) \tag{3}$$

The received signal r(t) in the kth symbol can be represented as

$$r(t) = \int h(t,\tau)y(t-\tau) = \sum_{l=0}^{\chi} h_l(t)y(t - \tau_l) + n(t) \tag{4}$$

n(t) is the background noise.

For practical implementation, modulation and demodulation can be achieved by Inverse Fast Fourier Transform (IFFT) and Fast Fourier Transform (FFT), respectively. Channel estimation is applied to obtain the estimates of channel fading in each sub-carrier such that coherent detection is achieved. Let q be the sub-carrier index at the output of the OFDM demodulator in a WiMAX system and $s_{k,q}$ is the output for the qth sub-carrier in the kth symbol interval. Now, let us assume that the channel impulse response is quasi-static during the kth symbol interval so that $h(t) \approx h(kT)$ for $kT \le t < (k+1)T$, the Inter Carrier Interference can be neglected compared to the background noise n(t). Thus the kth sub-carrier output, $s_{k,q}, q \in \{0, 1, 2, ..., M-1\}$, from the demodulator can be expressed as

$$s_{k,q} = \frac{1}{T}\int_{kT}^{(k+1)T}\left[\sum_{l=0}^{\chi}h_l(kT)\times\sum_{m=0}^{M-1}b_m[k]e^{j2\pi m(t-\tau_l)/T} + n(t)\right]e^{-j2\pi qt/T}$$

$$= \frac{1}{T}\sum_{l=0}^{\chi}h_l(kT)\sum_{m=0}^{M-1}b_m[k]e^{-j2\pi m\tau_l)/T}\times\int_{kT}^{(k+1)T}e^{j2\pi(m-q)t/T}dt + \frac{1}{T}\int_{kT}^{(k+1)T}n(t)e^{-j2\pi qt/T}dt \tag{5}$$

$$= c_{k,q}H_{k,q} + v_{k,q}$$

where,

$$c_{k,q} = b_q[k], H_{k,q} = \sum_{l=0}^{\chi}h_l(kT)e^{-j2\pi q\tau_l/T} \text{ and } v_{k,q} = \frac{1}{T}\int_{kT}^{(k+1)T}n(t)e^{-j2\pi qt/T}dt \tag{6}$$

If the channel fading is characterized by $H_{k,q}$ were known, then coherent detection and optimum diversity combining would be achievable at the receiver. $H_{k,q}$ is time varying and usually unknown. Hence, for the accurate estimation of the channel fading parameters, $H_{k,q}$ is to be evaluated for a given $s_{i,q}$, $i \le k$ and q = 0, 1, ..., M-1.

3   WiMax Simulation at baseband level

The full WiMAX simulation is shown in the Fig 3.

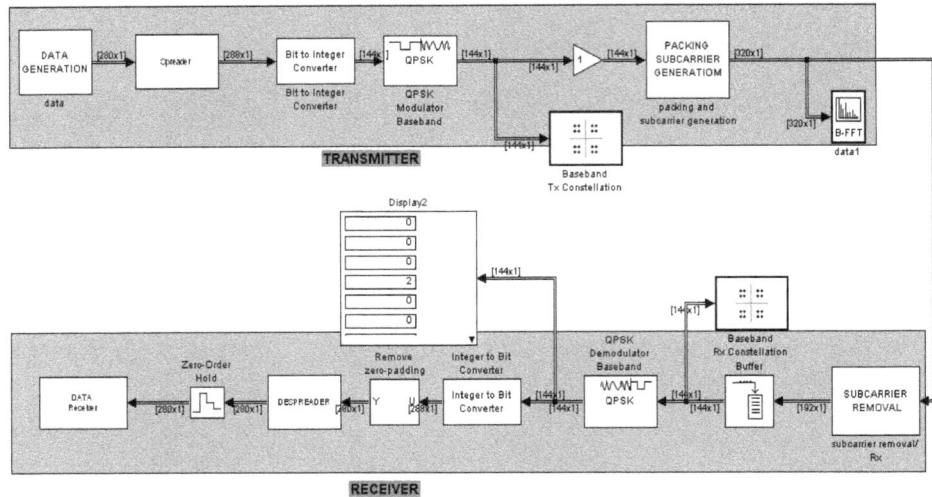

Fig. 3. The Simulink model of WiMAX transmitter and receiver.

The descriptions of some important blocks are as follows:

(1) Data generator [MAC PDU] block under 'data' block:

The 802.16 standard is designed with the network and communication security keeping in considerations. Over the wireless link the robust security aspects are to be considered as the most important to control the confidentiality during data communication. In 802.16 standards, the security keys and encryption techniques are involved as shown in the Fig. 4. It has the similarity in concepts of adopting the security parameters as of IPsec. After the authorization from Security Association, the X.509 certificate, consists of an authorization key (AK), a key encryption key (KEK), and a hash message authentication code (HMAC) key, which are used for authorization, authentication, and key management [6] Here in the following model, we have 10 blocks utilized for message authentication and security management. From the top to bottom, those blocks are: (i) HT: Header Type, (ii) EC: Encryption Control, (iii) Type: Payload Type, (iv) RSV: Reserved, (v) CI: CRC Identifier, (vi) EKS: Encryption Key Sequence, (vii) RSV: Reserved, (viii) LEN: Length of Packet, (ix) CID: Connection Identifier and (x) HCS: Header Check Sequence.

In terms of message authentication, there are some important shortcomings in IEEE 802.16 standard implemented at the MAC Layer. To avoid the serious threats arise from its authentication schemes, the WiMAX involves a two-way sequential transactions for controlling, authorization and authentication. During the basic and primary connection, MAC management messages are sent in plain text format which is not a robust type of authentication and so can be easily hijacked over the Air and this can be done by the attacker once again. So, as per the X.509 Certificate, the Public Key Infrastructure (PKI) defines a valid connection path to identify a genuine Security Systems. It uses RSA Encryption with SHA1 hashing. The certificate, as pre-configured by the specific manufacturer and embedded within the system must be kept secret so that it can not be stolen by other users/vendors. A Security System that is certificated by a particular manufacturer is implemented in a Base Station (BS) and the particular BS can not know the internal standards priorly.

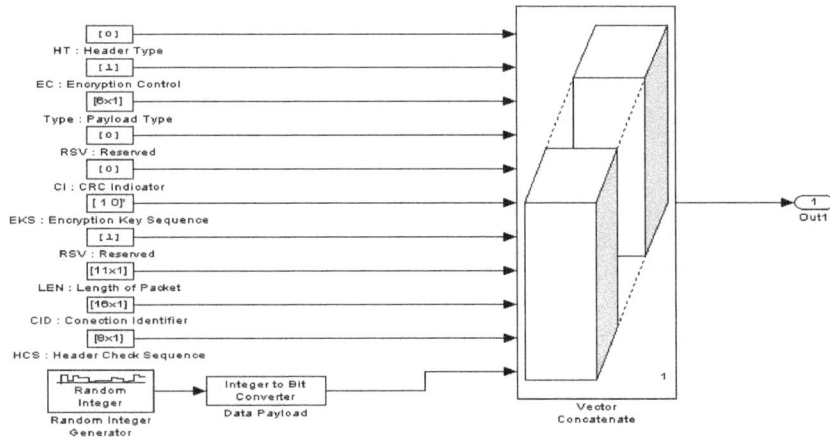

Fig. 4. The MAC PDU generator including header scheme and payload

Since, mutual authentication verifies the genuineness of a BS, it should be present in any wireless communication as it is virtually open to all. Extensible Authentication Protocol (EAP) is mostly utilized in any WiMAX Base Stations as to protect IEEE 802.16 / WiMAX against masquerading parties.

The spectrum of the base band just after the Mac PDU packing is shown in Fig.5.

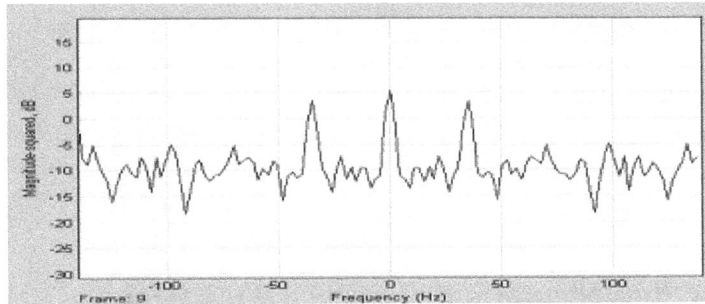

Fig. 5. Base band spectrum

(2) The PN spreading block

The incoming signal is XOR'd with the bit pattern generated by a PN Sequence Generator. This is further zero padded to increase the frame size to 288×1. The Chip sample time is 1/1000 S, so the chip rate is 1 KHz. The spectrum after spreading looks like as shown in the figure 6 below: The bit pattern generated after this is fed into a bit to integer converter and then to a QPSK modulator.

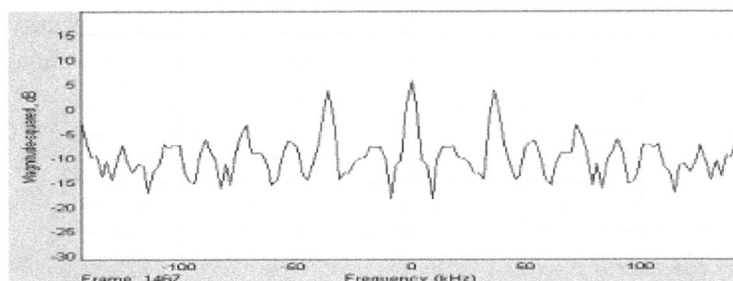

Fig. 6. Spectrum after spreading

(3) The Sub carrier Generation and packing sub system:

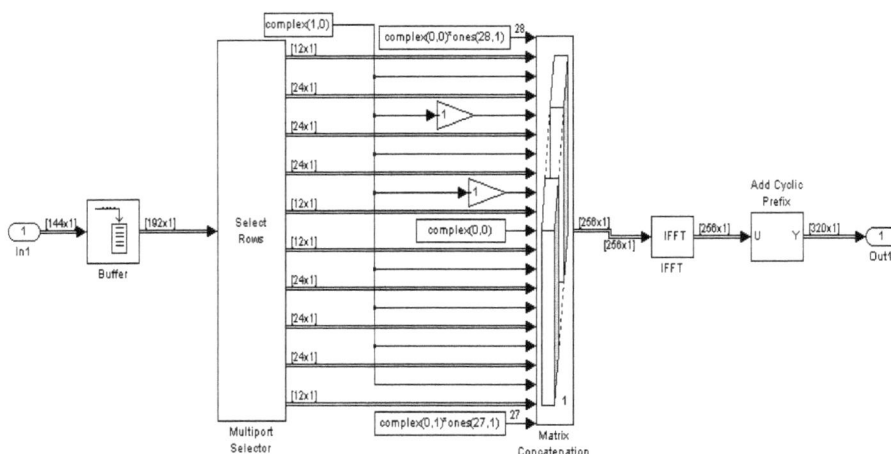

Fig. 7. Input packing before IFFT transformation and cyclic prefixing

After the QPSK modulation, pilot is inserted which helps in channel estimation. Here 192×1 input stream is broken down into 10 different data pipes and pilots inserted in between them according to the above figure. All the rows of the resulting data are combined before feeding it to the IFFT block for sub carrier generation in time domain and then cyclic prefixing to add guard time. The final Tx blocks looks like as depicted in Fig.7.

4   Hardware Implementation of the 60 GHz System

4.1   Description of Transmitter Section

The prototype model of the 60 GHz transmitter is shown in Fig.8 and its block schematic diagram is shown in Fig.9.The Tx section consists of several parts as shown in Fig.9. The PC is used to programme the VSG using Matlab/ Simulink for the generation of two orthogonal basis functions [9]. In the base band section we programmed the ARB section of the VSG to generate the base band WiMAX signal and it is then up converted to IF level of 1 GHz and fed to the 60 GHz varactor tuned Gunn oscillator. The basic block diagram is shown in the Fig.9. The Gunn oscillator is followed by 60 GHz attenuator and frequency meter for the control and frequency measurement of 60 GHz transmitted signal respectively [7]. The 2 feet parabolic disc antenna is connected at the output for radiation of 60 GHz signal

Fig. 8. The 60 GHz Transmitter

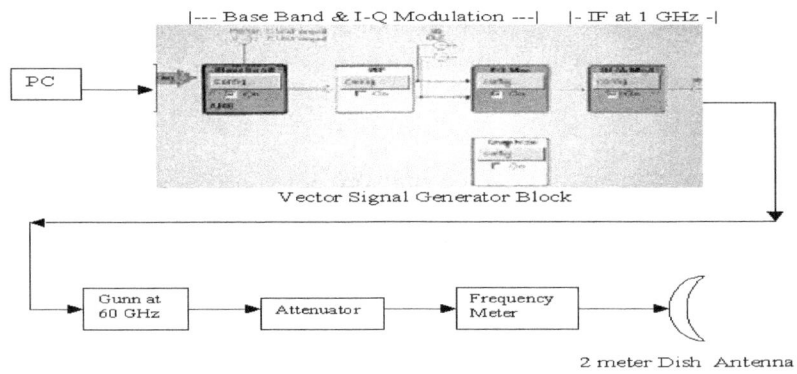

Fig. 9. Block diagram of the WiMAX Transmitter

4.2 Description of Receiver Section

The prototype model of the 60 GHz receiver is shown in Fig.10. The block schematic diagram is shown in Fig.11 where the receiver consists of a front end, which receives signal through a horn antenna. The received signal then down converted to IF level (1 GHz) using 61GHz Gunn oscillator. This signal is further amplified by two IF amplifiers and is fed to input of the DVB satellite receiver tuner. The I-Q signal from the receiver tuner is connected to the RTSA as shown in Fig.11. We store the received I-Q data in RTSA for further analysis, as shown in Fig.14.

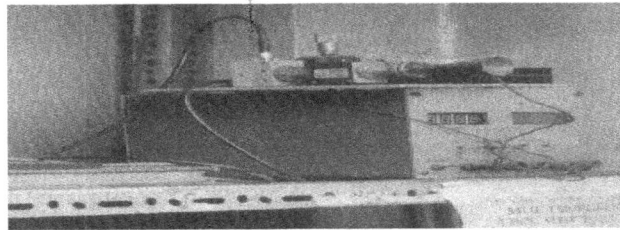

Fig. 10. The 60 GHz Received RF Front End

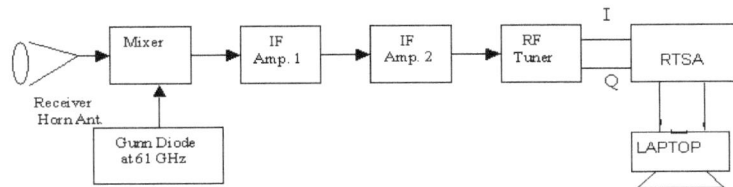

Fig. 11. Block diagram of the 60 GHz WiMAX receiver

The Rx spectrum at IF level is shown in Fig.13.

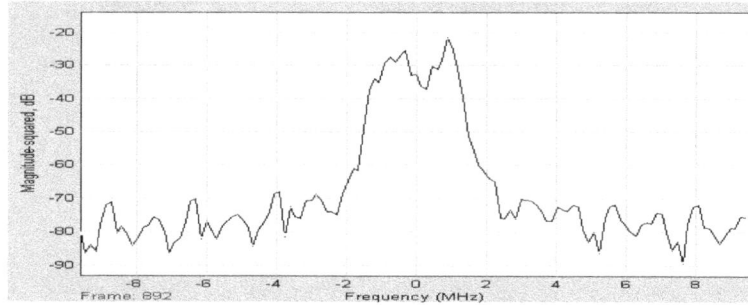

Fig. 12. The final transmit spectrum

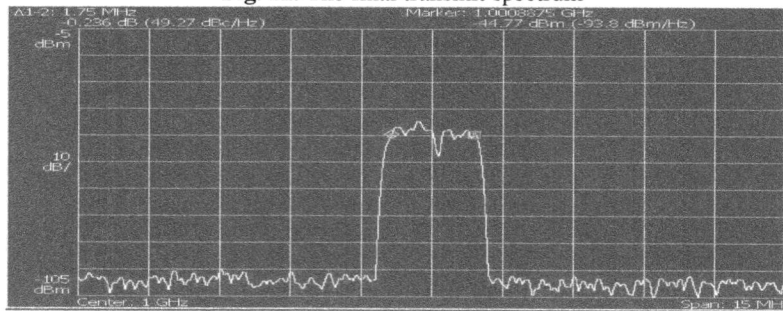

Fig. 13. Received spectrum with bandwidth is 1.75 MHz

The Received In-Phase and Quadrature-Phase Signals in Real Time Spectrum Analyzer is shown in figure 14.

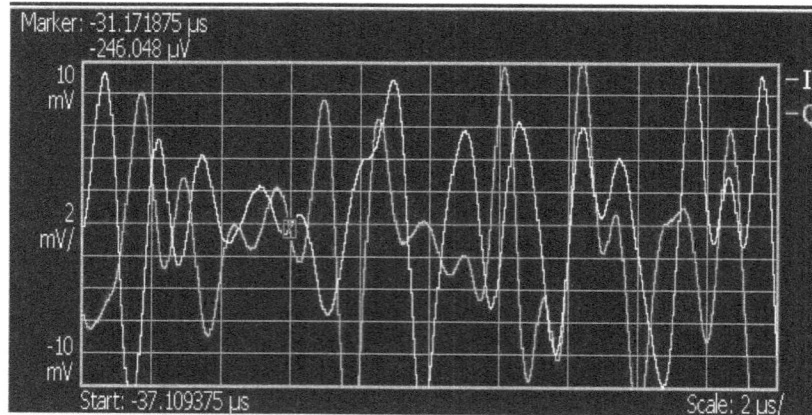

Fig. 14. Received I, Q Signals in RTSA

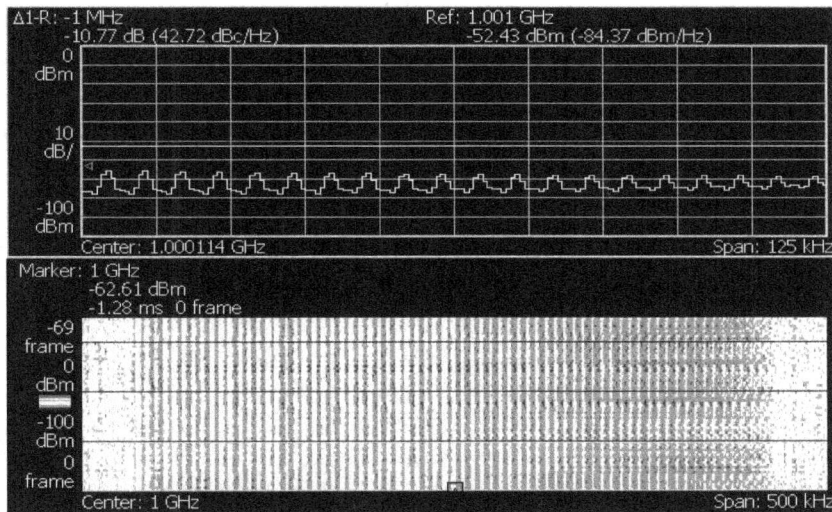

Fig. 15. Received WiMAX Sub-Carriers in RTSA

5   Conclusion

Lots of efforts are imparted for the development of the 60 GHz C2C link. The system is tested at SMIT laboratory with multimedia transmission and reception. Technical expertise are developed towards Simulink programming, methods of poring to VSG, IF and millimeter wave hardware, RTSA use, Data Acquisition and DSP. The system is operational at SMIT laboratory but yet to be tested after mounting on the vehicles. This successful development encourages the active groups at the laboratory. With proper deployment of this 60 GHz system on vehicles, the existing commercial products for 802.11P will be required to be replaced or updated soon and we look forward for the improved society with intelligent vehicles.

References

1. 60GHz Wireless Communications: Emerging Requirements and Design Recommendations Robert C. Daniels and Robert W. Heath, Jr.
2. WiMAX Security for Real-World Network Service Provider Deployments
3. White paper Airspan Mobile WiMax Security
4. DENSO's WSU Unit Car-to-Car Communication Consortium
5. C2C-CC Security Baselines, Car2car communication consortium. www.car-to-car.org
6. IEEE 802.16/WiMax Security Hyung-Joon Kim Dept. of Electrical and Computer Engineering Stevens Institute of Technology, Hoboken, New Jersey.
7. Millimeter-Wave Radar Technology for Automotive Application *by Shinichi Honma and Naohisa Uehara*
8. Channel Allocation and Routing in Wireless Mesh Networks: A survey and qualitative comparison between schemes F. Kabbi, S. Ghannay, and F. Filali International Journal of Wireless & Mobile Networks (IJWMN), Vol.2, No.1, February 2010
9. Multi-Standard Programmable Baseband Modulator For Next Generation Wireless Communication Indranil Hatai and Indrajit Chakrabarti, International Journal of Computer Networks & Communications (IJCNC), Vol.2, No.4, July 2010

3

EFFECTS OF NOVEL LONG LIFE ROUTING ALGORITHMS ON IEEE 802.16 THROUGHPUT IMPROVEMENT METHODS FOR VARIOUS VEHICULAR VELOCITIES

Barbaros Preveze

Department of Electronics and Communication Engineering, Çankaya University, Ankara, Turkey

b.preveze@cankaya.edu.tr

ABSTRACT

In this paper, effects of Fastest Path Routing, Ant-Colony Routing, Associativity Based Routing (ABR) and other novel proposed Associativity Based Long Life Routing algorithms, on novel proposed throughput improvement algorithm Most Congested Access First (MCAF), minimizing the packet loss ratio and improving the throughput of a multi-hop mobile WIMAX network are investigated for various vehicular velocities. The proposed Long Life Routing algorithms are shown to provide more throughput improvement for all vehicular velocities and provide longer life routes.

KEYWORDS

Throughput, Routing, 802.16j, Multimedia

1. INTRODUCTION

The idea of Cognitive Radio was first presented officially by Joseph Mitola III in a seminar at KTH, The Royal Institute of Technology, in 1998. Then an article is published [1] by Mitola and Gerald Q. Maguire, Jr one year later. Mitola introduced this research as a baby step in a potentially interesting research approach [1] and described Cognitivity later as; "The point in which wireless personal digital assistants (PDAs) and the related networks are sufficiently computationally intelligent about radio resources and related computer-to-computer communications to detect user communications needs as a function of use context, and to provide radio resources and wireless services most appropriate to those needs" [2]. By the appearance of this novel concept new challenges in the literature to the resource allocation and design of WIMAX relay-based system have been introduced. Most of the works in the literature attempt to improve the system throughput by trying to cooperate the primary (licensed) users and secondary (unlicensed) users for efficient resource allocation [3, 4]. However, there are only little research works specifically on multi-hop 802.16j networks at present [5] that try to improve the throughput with design of WIMAX relay based system.

In [3], the authors propose a novel method for flexible channel cooperation which allows secondary users to freely optimize the use of channels for transmitting primary data along with their own data in order to maximize the throughput performance.

In [4], the throughput potential of cognitive communication is explored considering opportunistic communication as a base line and the throughput improvements are offered by the overlay methods. The authors focus on determining the throughput potential of Cognitive Radio for various transmission powers of the secondary nodes and determining the optimal amount of licensing. However design of WIMAX relay-based system, effects of routing methods and effects of vehicular velocities are not considered in [3] and [4].

In [5], a study of the performance of transparent mode relay-based 802.16j systems is described and design of WIMAX relay-based system is considered. However effects of routing algorithms and effects of vehicular velocities are not considered in this work and only 5% throughput improvement is provided with almost twice signaling overhead where we have provided the throughput improvement for different number of nodes (N) in a range between 15% (for N=24) and 36% (for N=4) as seen in Fig.5. According to authors [5], the work already demonstrates the necessity of finding other ways to improve 802.16j network throughput performance.

In [6], the authors address the problem of assigning channels to Cognitive Radio transmissions assuming one transceiver per Cognitive Radio in order to maximize the number of simultaneous Cognitive Radio transmissions with respect to spectrum assignment. They provide 50% improvements on the network throughput for single hop scenarios by decreasing the blocking rates of Cognitive Radio transmissions. However effects of routing algorithms and effects of vehicular velocities are not considered in this work and only 20% improvement is provided for multi-hop scenarios, where we have provided the throughput improvements up to 36% (for N=4) as seen in Fig.5.

In our system we use the proposed OFDMCAF method which is a combination of OFDMA and TDMA [7, 8, 9]. It uses additional novel throughput improvement methods to decrease the packet loss ratio and improve the throughput of a unicast network.

In [10], simulation throughput results of each throughput improvement method are focused on, only for a fixed number of nodes (N=6). In [11], the simulation results are generalized and evaluated for varied number of nodes (N) and it is shown that the simulation results evaluated for pure network match with the results of another work [12] in the literature. In [11] the probabilistic calculation of packet loss rates of each method is also formulated and the evaluated packet loss rate calculation results are confirmed by the evaluated simulation results. The calculation of the bandwidth wastage appeared in [11] for the first time. However, in this work, the results evaluated by 3 different works in the literature [12, 13, 14] are used to confirm the results of the simulations and the calculations we have evaluated. The throughput calculation with/without bandwidth wastage is formulated and generalized for different N values using the changes in; number of nodes, average hop counts, packet loss rates and number of successfully sent/lost video/voice/data packets. Then the results of the calculations evaluated for pure network are confirmed by the simulation results of same cases for each N value. It is shown that the real time packet transmissions have been provided with full success, the packet loss rate of non-real time data packets has been minimized and the throughput has been improved by each novel method. Finally, as the main purpose of this paper, the effects of Fastest Path, Ant-Colony, Associativity Based Routing (ABR) and novel proposed Long life routing algorithms on novel throughput improvement methods are investigated. It is shown in this paper that, even all routing algorithms improve the system throughput; proposed ABR methods provide greater throughput improvement for all proposed throughput improvement methods for all vehicular velocities. But, the amounts of improvement decreases by increase of vehicular velocity for all cases.

2. THROUGHPUT OF MOBILE AD-HOC NETWORKS

In order to confirm the correctness of our simulation results, the pure system simulation results are compared with the results of the calculations or simulations ones evaluated in the literature [12, 13, 14] using the same parameter sets of [12],[14] and our simulation system. The re-simulated results of the works [12, 13, 14] for the parameters used our simulation are shown in Fig.1. Note that the results of [12] are evaluated using (1) [12], results of [13] are evaluated using (2) [13] and the results of [14] are evaluated using (3) [14]. Our results are consistent with [12, 13, 14].

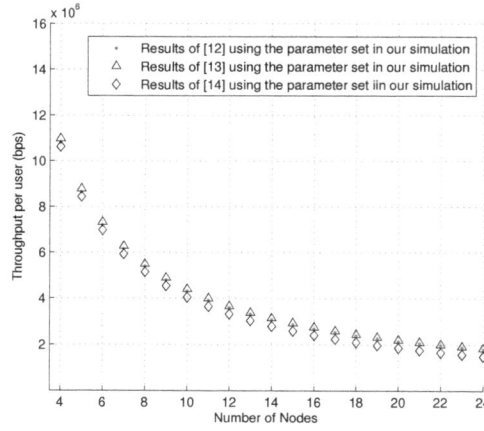

Fig. 1. Comparative analysis for throughput results of different works

2.1 Asymptotic throughput analysis of unicast transmission

One of the resultant formulations of the works in the literature that evaluates the throughput of a unicast system by asymptotic analysis is given in (1) [12]. It is also divided by the number of nodes to evaluate the throughput per user.

$$\text{Ru} \; \Box \; B \log_2 \left(1 + \frac{\rho_0 \ln(N)}{d_c^{\;n}} \right) + \frac{Bn}{2\ln(2)}$$

$$where \; \rho_0 = \frac{P}{N_0} \times K \times \beta$$

$$\text{(1)}$$

$$E\left\{ R_i^s \right\} = \frac{1}{N} \times E\left\{ R_{total}^{Hom} \right\}$$

$$= \frac{1}{N} \times B \times \log_2 \left(1 + \beta \times E\left\{ \Gamma_{eff} \right\} \right)$$

$$\text{(2)}$$

On the other hand, there is another asymptotic analysis done [13] that the throughput per user is formulated there as given by (2). "Γ_{eff}" in (2) is the average effective SNR. The results illustrated in Fig.1 are evaluated for all works using the same parameter values as in our simulation program (B = 10MHz; FL = 5ms; SNR = 103; dc = 50m; n = 2(freespace);K = (0dB); _ = 0:02). In order to confirm the correctness of our pure system simulation and calculation results, the results of [12] in Fig.1 which exactly match with the results of [13, 14] are carried on Fig.5 with the legend "Unicast analysis results of pure OFDMA without Adaptive Rate (AR), Buffer management (BM) and Spectral Aids (SA)". It's shown in Fig.5 that," simulation and calculation results of pure OFDMCAF without AR, BM and SA", match with "Unicast analysis results of pure OFDMA without Adaptive Rate (AR), Buffer management (BM) and Spectral Aids (SA)" especially for large number of nodes as depicted in [12]. It is also shown that, "Calculation results of pure OFDMA system without AR, BM and SA" match exactly with "Unicast analysis results of pure OFDMA without Adaptive Rate (AR), Buffer management (BM) and Spectral Aids (SA)".

2.2 WIMAX throughput evaluation of conventional relaying

The throughput evaluation of conventional relaying is also evaluated using (3) [14], with the case we have in our simulation that relay stations are in use and one can transmit at once.

$$BR = \frac{R\_OFDM(RS)}{LoF\left(\sum_{i=1}^{n} \frac{SSG1_i}{bps_i} + \sum_{i=1}^{n} \frac{SSG2_i}{bps_i} + \frac{1}{bps} \sum_{i=1}^{n} SSG2_i \right)}$$

(3)

3. SIMULATION PROGRAM

We have developed an event-driven simulation program using MATLAB, in which the movements, locations and buffer states of the nodes, the organization of the packets in the buffer of each node, selected routes for each packet, instant data generation/transmission rates, and instant overall throughput values are all observable from the screen during the simulation. In our simulation, N independent and uniformly distributed mobile relay nodes are considered to communicate with each other in a cell structure. The overall algorithm that the nodes use in the simulation is given in Fig.2 where any simulation parameter value can also be changed to any desired value.

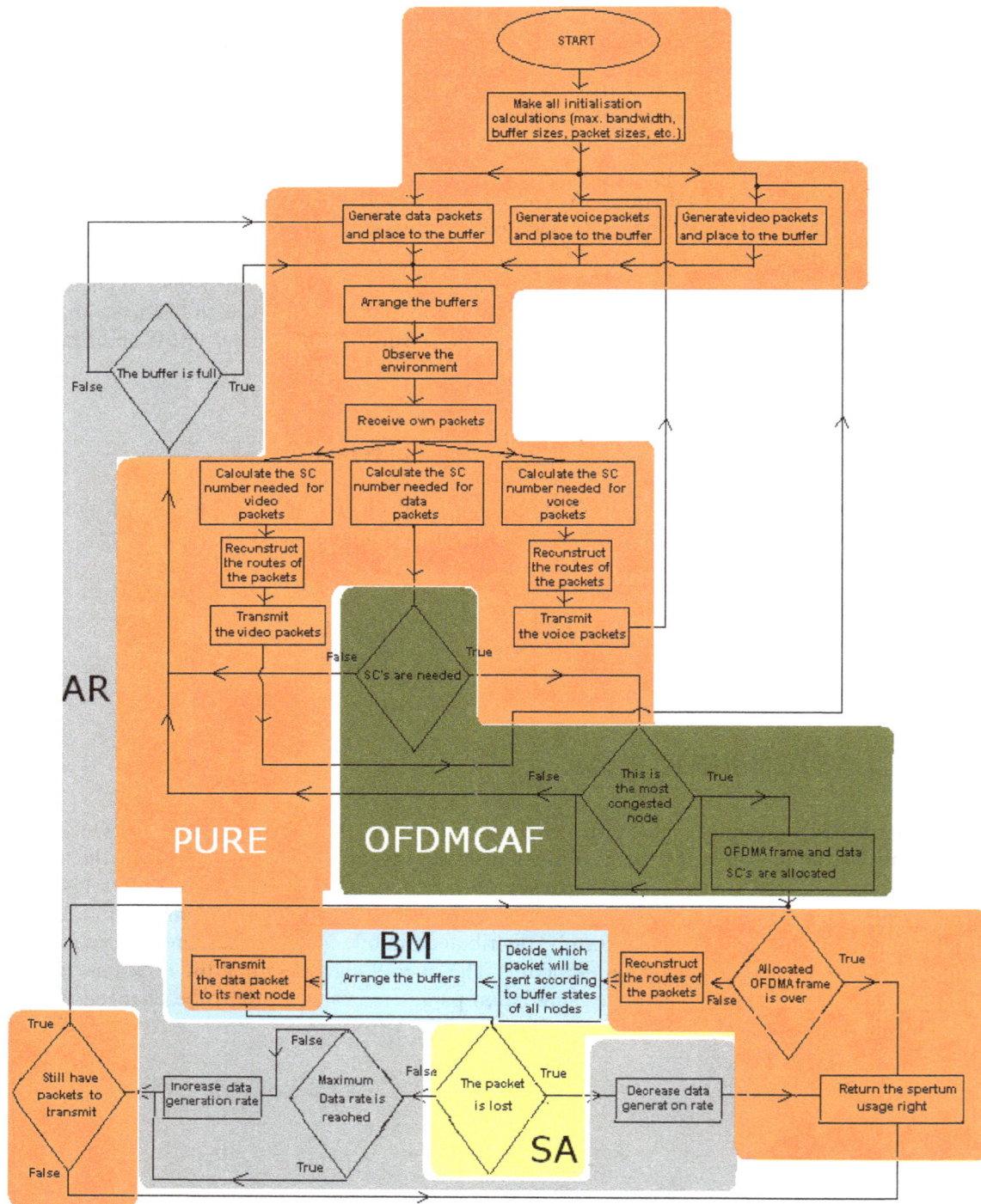

Fig. 2. The overall algorithm used by each node in the simulation in case of using OFDMCAF with AR, BM and SA It is also assumed for best effort that we have unlimited packet generation at data queues of the nodes for which the transmission rate changes according to network conditions.

3.1 Parameters used in the simulation and determination of the maximum spectral usage

In the simulation program the maximum spectral usage that a system may provide is determined using the given input parameters between (4) - (10).

$$Modulation\,(\text{QAM})\ =\ 64 \tag{4}$$

$$Frame\ Length\,(FL)\ =\ 5\ ms \tag{5}$$

$$Forward\ Error\ Correction\,(FEC)\ =\ 3/4 \tag{6}$$

$$Sub\text{-}channel\ Capacity\,(SCC)\ =\ 0.95\ Mbps \tag{7}$$

$$Data\ symbols\ per\ Frame\,(DSPF)\ =\ 44\ of\ 48 \tag{8}$$

$$Number\ of\ sub-channels\,(NOS)\ =\ 30 \tag{9}$$

$$Bandwidth\,(B)\ =\ 10\ MHz \tag{10}$$

$$Frames\ per\ second\,(FPS) = 1/FL\ f/s \tag{11}$$

$$Bits\ per\ symbol\,(BPS) = log_2\,(QAM)\times NODS\ b/sym. \tag{12}$$

where the abbreviation NODS in (12), expresses the number of data sub-carriers.

$$Bits\ per\ symbol\ with\ FEC\ (BPSWF) = FEC\times BPS\ b/sym \tag{13}$$

$$Minimum\ Allocatable\ Unit\,(MAU) = BPSWF/NOS\ bits \tag{14}$$

$$sub\text{-}channel\ data\ rate\ in\ a\ frame\ with\ FEC\ (SCDRFWF) = DSPF\times MAU\ b/f \tag{15}$$

$$Symbol\ rate\ per\ sec.\,(SRPS) = FPS\times DSPF\ sym./s \tag{16}$$

Using the size of Data Packet (DTPS) as 150 Bytes, Voice Packet size (VCPS) as 10 Bytes [7], Initial Reference Video Packet size (VDPSref) as 512 Bytes, and using other parameters as defined by the standards [7, 9] given in (4) - (10) the parameters are calculated using the equations in (11)- (16);

from (12), we have;

$$Bits\ per\ symbol\,(BPS) = 6\ bits\ /subcarrier\times 720\ subcarrier$$
$$= 4320\ b/sym \tag{17}$$

from eq. 13, we have;

$$Bits\ per\ symbol\ with\ FEC\ (BPSWF) = \frac{3}{4} \times 4320$$
$$= 3240\,b/sym.$$
(18)

For NOS = 30, B=10 MHz, using eq. 13, eq. 14 and the result of eq. 18 we have;

$$Minimum\ Allocatable\ Unit\ (MAU) = 3240 / 30$$
$$= 108\ bits$$
(19)

Considering a device is allotted one sub-channel (SC) in every frame, there will be 200 frames/sec (from eq. 11) with DSPF= 44 OFDM data symbols/frame each, so the symbol rate results with 8800 using eq. 16 as in eq. 20.

$$Symbol\ rate\ per\ sec.\,(SRPS) = (200 \times 44)$$
$$= 8800\ data\ sym./s.$$
(20)

As the resultant value is also given in eq. 7, the value of sub-channel capacity with FEC can be calculated [7] by eq. 21,

$$Sub\text{-}channel\ capacity\ with\ FEC\ (SCC) = (MAU \times SRPS))$$
$$= 108\ bits \times 8800\ sym./s.$$
$$= 950400\,bps.$$
(21)

Then, the maximum spectrum usage (MSU) can be calculated as in eq. 22, by multiplying the sub-channel capacity with the number of sub-channels.

$$Max.\ spectrum\ usage(MSU) = SCC \times NOS$$
$$= 950400\ bps \times 30$$
$$= 28.512 \times 10^6\ bps$$
$$= 3564000\ Bps$$
(22)

The calculated result of eq. 22 also matches with the other results evaluated in the literature [7, 9].

3.2 Determination of maximum spectral usage with/without bandwidth wastage

In [10, 11] the bandwidth wastage of voice packets in a second (Wastage_voice) for a given N value is calculated by (23).

$$Wastage_{N_{voice}} = \left((N \times \frac{MSU}{NOS \times \frac{1}{FL}}) - \left(VCPS\ x\ (N + (N\text{-}1)\ x\ N)\ x\ VPSR_{voice}\right) \right) \times \frac{1}{FL}\,Bps$$
(23)

Then the maximum possible successful spectral usage (MSU) taking account the bandwidth wastage is calculated as in (24) by using the given parameters and by subtracting the total bandwidth wastage amount calculated in (23) from maximum spectral usage.

$$MSU_{with\_wastage} = MSU - Wastage_{N_{voice}} \; Bps \tag{24}$$

The provided throughput amount not taking account (without) the bandwidth wastage is also calculated in [10, 11] as given in (25) where AHC expresses the average hop count.

$$THR_{without\_wastage_N} \; ; \; \frac{AHC_N \times 8 \times (THR_{N_{with\_wastage}} + \left(wastage_{N_{voice}}\right))}{N} bps \tag{25}$$

By a simple check for N = 6 and for "no method is active" case, using the average hop count (AHC) retrieved from the fastest path routing simulations for N = 6 in the given area (AHC = 3), we have the resultant throughput per user without bandwidth wastage as in (26) where bandwidth wastage amount is calculated as 640800 Bytes from (23) and applied to (25) (MSU = 3564000 Bytes; NOS = 30 SC; FL = 5 ms; V CPS = 10 Bytes; VPSR$_{voice}$ = 1).

$$THR_{without\_wastage_N} = \frac{3 \times 8 \times (1225488 + (640800))}{6} bps \tag{26}$$
$$= 7465152 \, bps$$

3.3 Calculation of the packet sizes of multi-hop mobile WIMAX

3.3.1 Calculation of real time multimedia packet sizes

The nodes can use any voice/video packet sending rates (VPSR $_{voice\backslash video}$) and the video packet size (VDPS) is calculated to be equal to capacity of a single sub-channel in a frame by dividing the maximum spectral usage to the number of sub-channels and number of frames in a second. Since we use the FL=5 ms and NOS=30 in our system, the video packet size can be calculated as in (27). According to [7], 16 kbps voice packets can be considered due to low latency requirements and the voice packet size is calculated as in (28), where our system has 200 frames in a second with frame length of 5 ms.

$$VDPS = SCDRFWF$$
$$= \frac{\dfrac{MSU}{NOS}}{\dfrac{1}{FL}} = \frac{\dfrac{3564000 \, bytes \, / \, s}{30 \, SC}}{200 \, fps} = 594 \, Bps \tag{27}$$

$$VCPS = \frac{16 \, kbits}{200} = 10 \, Bytes \tag{28}$$

3.3.2 Calculation of non-real time data packet sizes

Number of sub-channels in a frame not used by video/voice packets and allocated for data transmission can be calculated as in (29) where TSCFDT expresses the total number of sub-channels allocated for data packets, TSCFVD expresses the total number of sub-channels allocated for video packets and TSCFVC expresses the total number of sub-channels allocated for voice packets.

$$TSCFDT = (NOS - (TSCFVD + TSCFVC)) \qquad (29)$$

The chosen data packet size (DTPS) closest to initially set one ($DTPS_{ref}$) is calculated as in (30) such that multiple of the resultant packet size exactly fit to the sub-channel without any bandwidth wastage where sub-channel data rate in a frame with FEC (SCDRFWF) is calculated as in (27). Note that for $VDPS_{ref} = 512$ $Bytes$, we have the resultant video packet size (VDPS) as 594 Bytes using (30) for no bandwidth wastage by video packets.

$$DTPS = \frac{(SCDRFWF)}{floor\left(\dfrac{SCDRFWF}{DTPS_{ref}}\right)}$$
$$= \frac{594}{floor\left(\dfrac{594}{150}\right)} = 198\ Bytes \qquad (30)$$

3.4 Calculation of the buffer sizes

The calculated data buffer size (CDBS) for data and video/voice buffer sizes evaluated for our system are formulated in [10, 11] by multiplication of packet size and the average number of corresponding packet type in the buffers of all nodes in the network as shown in (31),(32).

$$CDBS = (SCDRFWF) \qquad \times (number\ of\ allocated\ SC\ for\ data) \qquad \times (AHC+1)$$
$$= (DSPF \times MAU \div 8) \times (NOS - N \times VSPR_{voice} - N \times VSPR_{video}) \times (AHC+1)\ Bytes \qquad (31)$$

$$calculated_{video\ /voice}\ Buffer\ Size = (packet\ size_{video/voice}) \times (VPSR_{video/voice}) \times (N) \times (AHC+1)\ Bytes \qquad (32)$$
$$with\ (VPSR_{video/voice}) \times (N) \times (AHC+1)\ slots$$

3.5 Working principles of the simulation

The nodes in our system can have any kind of updated information about other nodes in the network. This information is used with novel methods to improve the throughput of the network by decreasing the packet loss ratio given in (33).

$$Packet\ loss\ ratio = lost\ packets / sent\ packets$$
$$= lost\ packets / (lost\ packets + successfully\ sent\ packets) \tag{33}$$

Since the nodes in the network use random way point mobility model [15] with random velocities in the given velocity ranges, it is difficult to predict the complete routes that the packets should follow. So each node recalculates and determines the route and next node of each packet before forwarding it. When congestion and packet losses occur in the network, the transmission rate is decreased using adaptive data transmission method. It is shown in the literature [16] that adaptive rate also has a positive effect on the system throughput. Maximum data rate per user (MDRPU) in a frame is evaluated in [10, 11] as given in (34).

$$MDRPU = MSU\ (bytes/s) \times \frac{TSCFDT}{NOS} \times \frac{1}{N} \times (FL) \times \frac{Succesfully\ sent\ p.}{Succesfully\ sent\ p. + lost\ p.} Bps \tag{34}$$

The overall algorithm shown in Fig.2 is also used for evaluating the simulation results of [10, 11] in which only the Fastest Path Routing Algorithm is used to decide the routes of the packets. However, in this study, effects of Fastest Path, Ant-Colony, Associativity Based Routing (ABR) and our novel proposed Associativity Based Long life Routing algorithms on novel throughput improvement methods are investigated for different vehicular velocities.

If a node starts to send its packets, it arranges its buffer such that the packets with same next nodes are grouped to be sent together and the packet group whose next node has more free memory will be sent first according to BM method. This process continues during the current OFDMA frame as long as the transmitting node still has packets to send and the frame duration is not over. The role of BM method can be observed in Fig.2.

According to SA algorithm, when a node starts to transmit packets, if the buffer of the freest next node is also full, the transmitter loses its first trial packet of chosen destination and even if the frame duration is not over, it returns its spectrum usage right to the node who gave alert of the 'spectrum need' most. At the end of the frame, the spectrum will again be allocated to another node with most spectral need. The role of SA method is also highlighted in Fig.2.

3.6 Formulations of packet loss rates and system throughput per user

3.6.1 Calculation of the packet loss rates

In [11], the average packet loss probability at one of the remaining N-1 nodes (excluding the current transmitter and the packets it has) is calculated using (35) - (36). After calculating the probability of loosing a packet in $Node_n$ using (35), the probability of a packet to get lost in any of the nodes can be evaluated as in (36).

$$P_{lost}(n) = \frac{\dfrac{Packet\ Distribution\ rate\ of\ Node_n}{Sum\ of\ Distribution\ rates\ of\ all\ nodes} \times Total\ Packet\ count}{Buffer\ size}$$

$$= \frac{\dfrac{(n)}{N \times (N+1)}}{Buffer\ size} \times Total\ Packet\ count$$

$$P_{lost}(n) = \frac{2 \times (n) \times Total\ Packet\ count}{Buffer\ size \times N \times (N+1)} \tag{35}$$

$$P_{loss} = \frac{1}{N-1} \times P_{lost}(1) + \frac{1}{N-1} \times P_{lost}(2) + ... + \frac{1}{N-1} \times P_{lost}(N-1)$$

$$P_{loss} = \frac{1}{(N-1)} \left(\sum_{n=1}^{N-1} P_{lost}(n) \right) \tag{36}$$

Note that the term $P_{lost}(n)$ used in (36) - (39) is the one evaluated in (35) and the term P_{loss} in (36) - (39) expresses the probability of loosing a packet with used method(s) at any of the nodes. When all methods are activated in our system, the packet loss occurs if and only if all the packets in the buffer of the transmitting node have the next nodes with full buffers. When proposed Buffer Management method is activated, the packet loss rate is calculated as given in (37).

$$P_{loss\_BM} = \frac{\binom{N-1}{1}P_{lost}^{1} + \binom{N-1}{2}P_{lost}^{2} + \binom{N-1}{3}P_{lost}^{3} + ... + \binom{N-1}{N-1}P_{lost}^{N-1}}{\sum\limits_{n=1}^{N-1}\binom{N-1}{n}} \tag{37}$$

$$P_{loss\_BM} = \frac{\sum\limits_{n=1}^{N-1}\left\{ \binom{N-1}{n} \times P_{lost}^{n} \right\}}{\sum\limits_{n=1}^{N-1}\binom{N-1}{n}}$$

The packet loss rate for the case of applying Spectral Aids method (P_{loss_SA}) without buffer management method (All - Buffer Management in Table 1), is calculated in [11] taking account the buffer state combination of remaining N-1 nodes, by sum of probability of choosing a full buffer and loosing the packet there.

So, for remaining N-1 nodes (excluding the transmitter itself), we have;

$$P_{loss\_SA} = \left(P_{lost}^{1} \times (1 - p_{lost})^{(N-1)-1} \times \frac{1}{N-1} \right) + \left(P_{lost}^{2} \times (1 - p_{lost})^{(N-1)-2} \times \frac{2}{N-1} \right) + ... + \left(P_{lost}^{n} \times (1 - p_{lost})^{(N-1)-(n)} \times \frac{n}{N-1} \right) \tag{38}$$

$$P_{loss\_SA} = \sum_{n=1}^{N-1} P_{lost}^{n} \times (1 - p_{lost})^{((N-1)-n)} \times \frac{n}{N-1}$$

When both BM and SA methods are both applied to the system, (39) is evaluated by combining (37) and (38) where confirmations of all evaluated calculation and simulation results are done in [11].

$$P_{loss\_rate\_ALL} = \sum_{n=1}^{N-1} \left(\frac{\binom{N-1}{n} \times p_{lost}^{n}}{\sum_{r=1}^{N-1}\binom{N-1}{r}} \times (1-p_{lost})^{((N-1)-n)} \times \frac{n}{N-1} \right) \tag{39}$$

3.6.2 Throughput calculation taking the bandwidth wastage into account

The average probabilities of losing packets at any node of determined routes are calculated using (40) and (41) respectively in case of using OFDMA and OFDMCAF [11].

$$p_{OFDMA_{ave}} = \frac{\dfrac{Total\ Packets}{N}}{Buffer\ Size} = \frac{Total\ Packets}{N \times Buffer\ Size} \tag{40}$$

$$p_{OFDMCAF_{ave}} = \frac{\left(\sum_{n=1}^{N-1}(n) \times \dfrac{2 \times Total\ Packets}{N \times (N+1)} \right)}{(N-1) \times Buffer\ Size} \tag{41}$$

The generalized throughput values for OFDMA and OFDMCAF can also be calculated for different N values. For this calculation we need; the simulation results only for a fixed N value, the calculations used in (40) and (41) and set of average hop count values for different N values evaluated from the routing simulation programs we additionally developed. So, the throughput per user for different N values with bandwidth wastage can be calculated using (42) and (43) by taking into account the effects of changes on packet loss rates, average hop count, number of nodes, value of V PSR$_{video}$, value of V PSR$_{voice}$ and number of remaining sub-channels for data packets by varying N value.

$$THR_{with\_BW\_wastage}(n) = \frac{\dfrac{AHC_{N_{sim}}\ for\ N\ used\ in\ Sim.}{AHC_{n_{sim}}\ for\ n\ nodes} \times [Transmitted\ packets(bytes)\ with\ corresponding\ packet\ loss\ rate]}{Time_{sim}(sec)}\ Bps$$

$$= \frac{\dfrac{AHC_{N_{sim}}\ for\ N_{sim.}}{AHC_{n_{sim}}\ for\ n\ nodes} \times \left[\dfrac{n}{N_{sim}} \times ((sent\ video)+(sent\ voice)) + \left((sent+lost\ data) \times \dfrac{TSCFDT}{TSCFDT_{sim}} \times (1-P(n)) \right) \right]}{Time_{sim}(sec)}\ Bps \tag{42}$$

$$\frac{Total\ Subchannels\ for\ Data\ Packets_{n}\ (TSCFDT)}{Total\ Subchannels\ for\ Data\ Packets_{SIM}\ (TSCFDT_{SIM})} = \frac{NOS - (n \times VPSR_{voice}) - (n \times VPSR_{video})}{NOS_{SIM} - (N_{SIM} \times VPSR_{voice_{SIM}}) - (N_{SIM} \times VPSR_{video_{SIM}})} \tag{43}$$

The abbreviation"P" used in (42) is taken as it is calculated in (40) or (41) depending on using it for calculation of THR$_{OFDMA}$ or THR$_{OFDMCAF}$. It's seen from Fig.5 that "Calculation results of pure OFDMCAF without AR, BM and SA" match with "Simulation results of pure OFDMCAF without AR, BM and SA" and "Calculation results of pure OFDMA without AR, BM and SA" also match with "Unicast analysis results of pure OFDMA without AR, BM and SA" evaluated in the literature [12, 13]. The calculated throughput results for activation cases of each method are also shown in Fig.5, for which all the calculation results match with the corresponding simulation results. Since mentioned works in the literature don't consider the bandwidth wastage on throughput calculation, we have also made the"throughput calculation without bandwidth wastage" to provide the fair comparison on Fig.5. The results in Fig.5 are presented for each number of nodes, with corresponding average hop count, video/voice packet sending rate ($VPSR_{voice}$ and $VPSR_{video}$) and other parameter values of the simulation at that instant.

3.7 Effects of routing algorithms on system throughput

As it is seen on both (25) and (42) that, the throughput amount is strongly dependent on AHC value such that the throughput is improved by decrease of the AHC value. In addition, according to (31) and (32), each node will need smaller buffers with smaller AHC values. Thus, the routes and hop counts determined by the routing algorithms have very critical role on throughput which has to be investigated. For this purpose, we have developed the routing simulation programs in MATLAB, implementing the Fastest Path routing [17], ANT- Colony routing [18], Associativity based routing (ABR), Alternative Enhanced Associativity based routing (AEABR) [19] and Associatively Tick Averaged Associativity based routing [20] algorithms. At first, effects of Fastest Path and ANT- Colony routing algorithms were taken into comparison for the range of different N values. The evaluated results are shown in Fig.6 where it is observed that Fastest Path routing algorithm reaches to higher throughput values. Then, Fastest Path routing algorithm and all other novel long life routing algorithms [19, 20] are embedded in our simulation system to run simultaneously and make their own decisions for the same conditions. Then, the simulations are run by activating different combinations of novel throughput improvement methods with different vehicular velocities for a fixed N value (N=6). Finally, the effects of routing algorithms on system throughput performance with different vehicular velocities are investigated for each of our novel throughput improvement methods. The evaluated results for different vehicular velocities and different routing algorithms are shown in Table 2 and Fig.7.

4. RESULTS and DISCUSSIONS

At each simulation run, the improvements made on; efficient bandwidth use, data packet loss rate and throughput of the system are evaluated with and without applying the novel methods for different number of nodes. By not taking one of the methods into progress, we have been able to observe the improvement and the effects of each method on system performance. It is depicted [5] that, the system throughput increase stabilizes when 4 relay nodes are deployed. Thus, before investigating all the results evaluated for a range of all N values, we focus on an example of results evaluated for totally N=6 nodes including the transmitter and the receiver [11]. Then, the simulation results are confirmed by comparing with other results evaluated in the literature and comparing by the results of calculations we made for packet loss rate and pure system throughput values. For better understanding; Fig.3 and Fig.4 are given to illustrate two examples of our simulation results for application cases of no methods and all methods respectively.

(a)

(b)

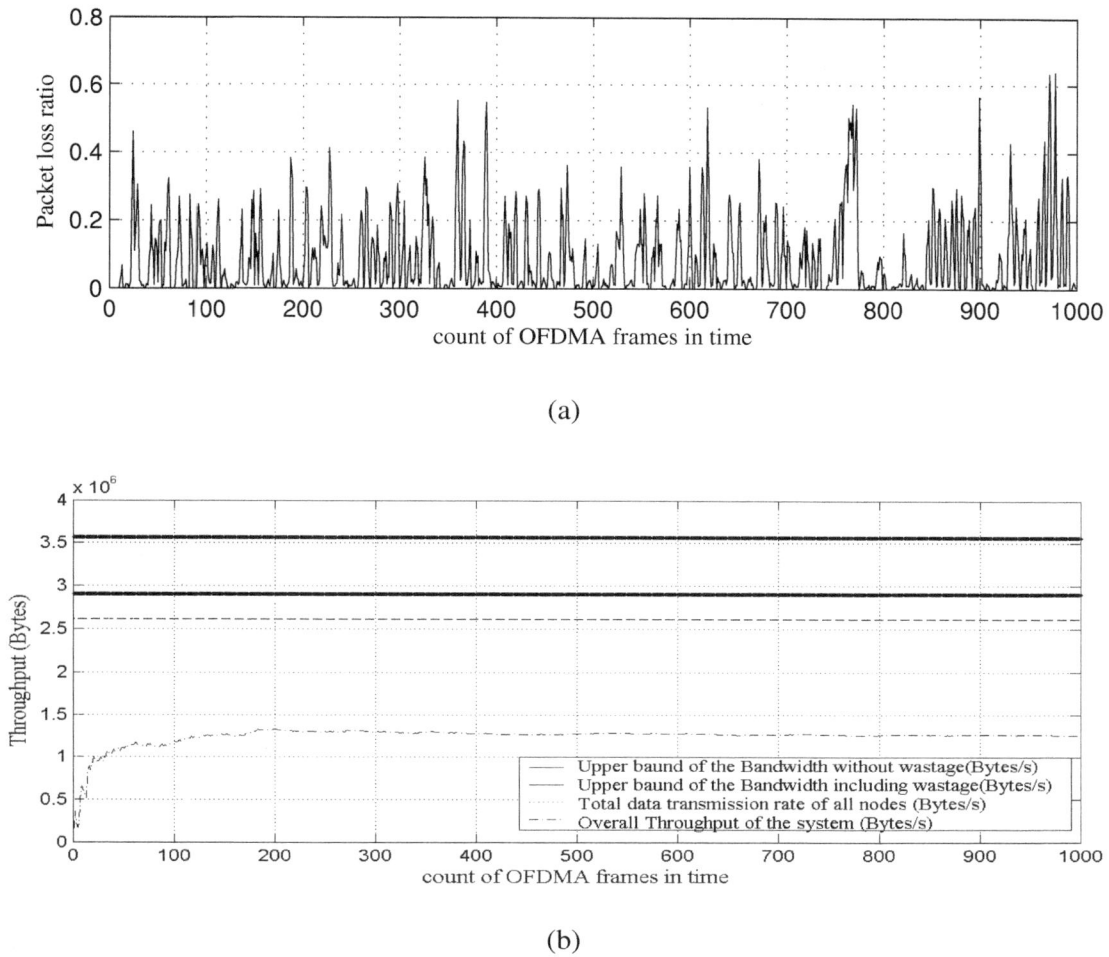

Fig. 3. (a) Packet loss Ratio of the whole network in last AHC frames period (b) Bandwidth upper bound, data packet generation/ transmission rate and overall throughput of the pure system vs. frame count, with use of half of the calculated data buffer size

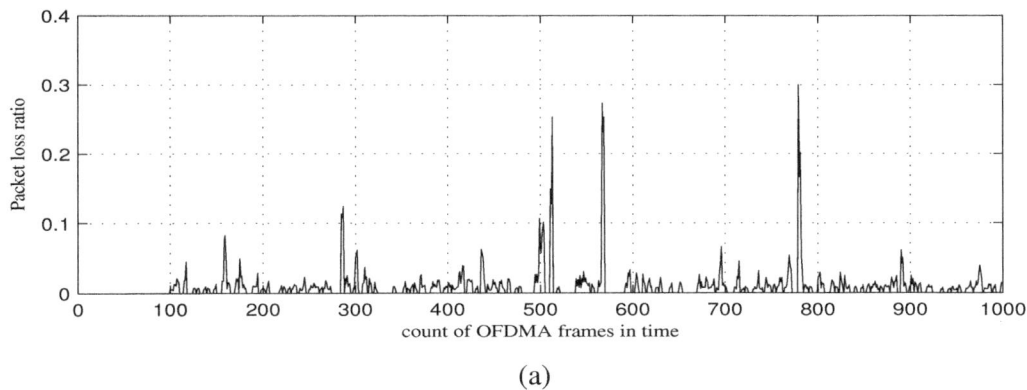

(a)

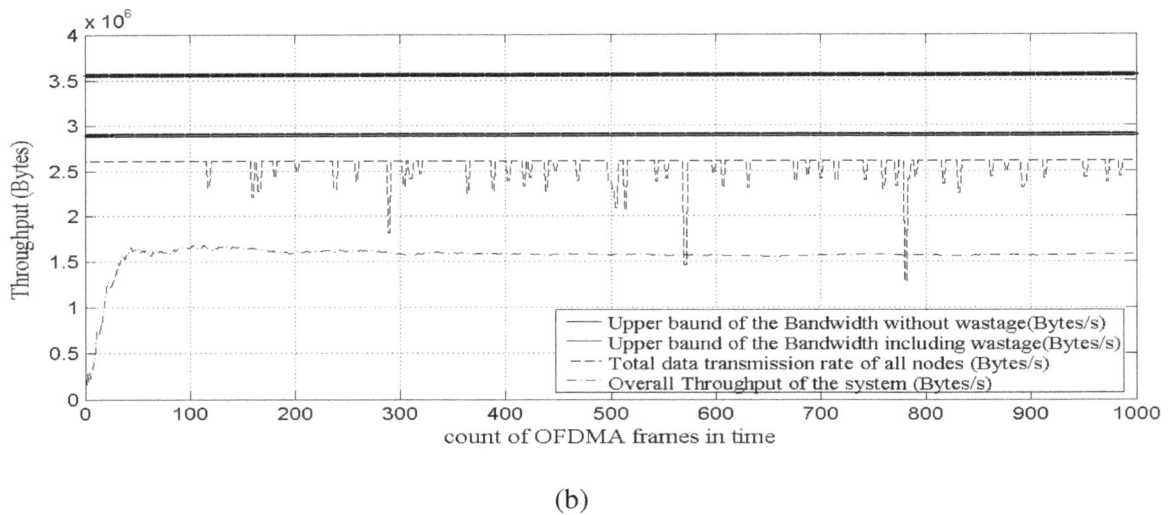

(b)

Fig. 4. (a) Packet loss Ratio of the whole network in last AHC frames period (b) Bandwidth upper bound, data packet generation/ transmission rate and overall throughput of the system vs. frame count, with use of all novel cognitive methods and calculated data buffer size

One example numeric values set of sent/lost packets [11], spectral usage rates and throughput amounts taken from the simulation results for each method are listed in Table 1 (for N=6), where the simulation throughput results for a range of different N values are also available on Fig.5.

Since the system throughput performance is reduced most during the absence case of spectral aids method (see Table 1), it can be said that the greatest throughput improvement is provided by spectral aids method among others. It is also seen that the most efficient spectrum usage is provided when all proposed methods are activated. Increasing the number of nodes also increases the number of total used sub-channels for real time video and voice packets, so, the number of remaining sub-channels to be allocated for data packets decreases.

Thus, the positive effects of our methods on loss ratio of data packets get weaker by increased number of nodes. As a result, it is observed in Fig.5 that, the throughput curves approach each other by increasing N value. We use 30 sub-channels in our system. There are N nodes sharing these sub-channels for transmitting their data, voice and video packets. We assume that each node uses, at least one separated sub-channel in one of every 4 frames [7] (with $VPSR_{video} >= 1/4$) for its own video conversations and at least one separated sub-channel in every frame [7] (with $VPSR_{voice} >= 1$) for its own voice conversations.

Thus, $[NOS - (N \times VPSR_{voice} + N \times VPSR_{video}))]$ sub channels will be used by N nodes for data packets, so we have $[NOS-(N+(N/4))] >= 0$ and $N <= 24$ *for NOS* = 30. That's why the value of N is increased up to 24 in our simulation (see Fig. 5).

Table 1. The simulation Results for different methods

Simulation Output Data in 5 seconds (for N=6)	Number of Video Packets		Number of Voice Packets		Number of Data Packets		Packet Loss Ratio		Effective Bandwidth Usage (Mbps) / %	Overall Throughput (Mbps)	Improvement
	Sent	Lost	Sent	Lost	Sent	Lost	Video/ Voice	Data			
No methods	494	0	2389	0	29344	4456	0%	13.18 %	2.64 / 90	1.22	0 %
All – adaptive	507	0	2576	0	35953	653	0%	1.78 %	2.79 / 95	1.48	22 %
All + larger	515	0	2532	0	36691	58	0%	0.16 %	2.80 / 96	1.51	24 %
All – buffer	517	0	2582	0	33978	654	0%	1.89 %	2.75 / 94	1.41	15 %
All – spectral	502	0	2575	0	31419	786	0%	2.44 %	2.78 / 95	1.30	7 %
All + Calc. B	515	0	2544	0	37631	182	0%	0.48 %	2.80 / 96	1.55	27 %

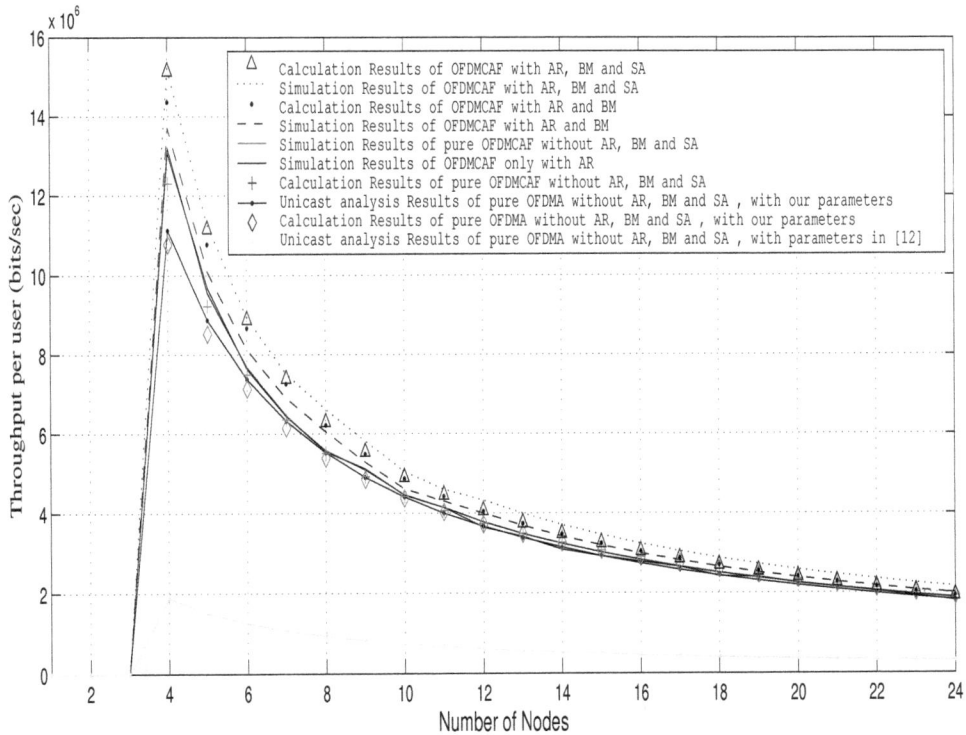

Fig. 5. Throughput results of OFDMA, OFDMCAF and unicast asymptotic analysis [12, 13] using fastest path routing algorithm

Before investigating the effects of all routing algorithms on the system throughput improvement methods, as the final purpose of this work, we investigated the effects of only the Fastest Path and ANT-Colony routing algorithms on throughput improvement of our methods with various numbers of nodes. It is shown in Fig.6 that the novel methods provide throughput improvement for both routing algorithms. But, it is seen that the system always provides higher throughput for the pure system with Fastest Path routing algorithm.

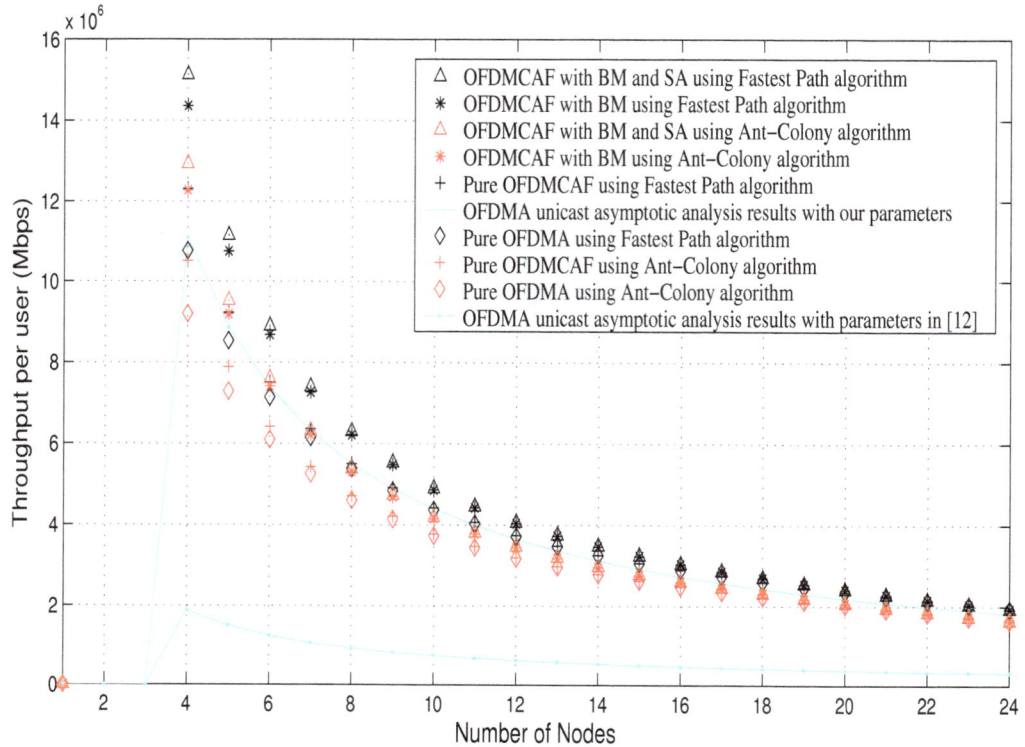

Fig. 6. The throughput performances of Ant-Colony and Fastest path routing algorithms

Finally, effects of Fastest Path routing algorithm (which is shown to always provide better performance than Ant Colony Routing algorithm) and Long Life routing algorithms (ABR, AEABR and ATAABR) on each novel proposed throughput improvement method are investigated in terms of instant and average; hop counts and system throughput value for different vehicular velocities in a mobile 802.16j network.

By running the simulator and processing all routing algorithms simultaneously for absolutely same conditions, the generated simulation results of all methods for different vehicular velocities are illustrated in Table 2. It is seen from the simulation results in Table 2 and Fig.7 for different vehicular velocity ranges that; the throughput is improved by ABR more than improved by Fastest Path routing algorithm (which is shown to provide greater throughput improvement than Ant-Colony routing algorithm in Fig.6). On the other hand the throughput is improved more by ATAABR then ABR where the throughput is improved most by our novel AEABR algorithm for all vehicular velocities. It is observed that, higher node velocities cause the throughput to decrease but the throughput amount decreases most for the vehicular velocities greater than 60 km /h.

Table 2. Average hop counts and throughput values of routing Algorithms for vehicular velocities between 25-40 km/h, 60-80 km/h and 100-110 km/h

Routing Method		25-40 km / h vehicular velocity				60-80 km / h vehicular velocity				100-110 km / h vehicular velocity			
Fastest Path Algorithm	Hop Count	3.16				3.16				3.16			
	Throughput (Mbps)	NO	BM	SA	BM,SA	NO	BM	SA	BM,SA	NO	BM	SA	BM,SA
		1.22	1.30	1.32	1.33	1.22	1.30	1.32	1.33	1.22	1.30	1.32	1.33
ABR Algorithm	Hop Count	2.12				2.16				2.20			
	Throughput (Mbps)	NO	BM	SA	BM,SA	NO	BM	SA	BM,SA	NO	BM	SA	BM,SA
		1.82	1.94	1.97	1.98	1.79	1.90	1.93	1.95	1.75	1.87	1.90	1.91
ATAABR Algorithm	Hop Count	2.10				2.20				2.13			
	Throughput (Mbps)	NO	BM	SA	BM,SA	NO	BM	SA	BM,SA	NO	BM	SA	BM,SA
		1.84	1.96	1.99	2.00	1.75	1.87	1.90	1.91	1.81	1.93	1.96	1.97
AEABR Algorithm	Hop Count	2.02				2.09				2.08			
	Throughput (Mbps)	NO	BM	SA	BM,SA	NO	BM	SA	BM,SA	NO	BM	SA	BM,SA
		1.91	2.03	2.07	2.08	1.85	1.96	2.00	2.01	1.85	1.97	2.01	2.02

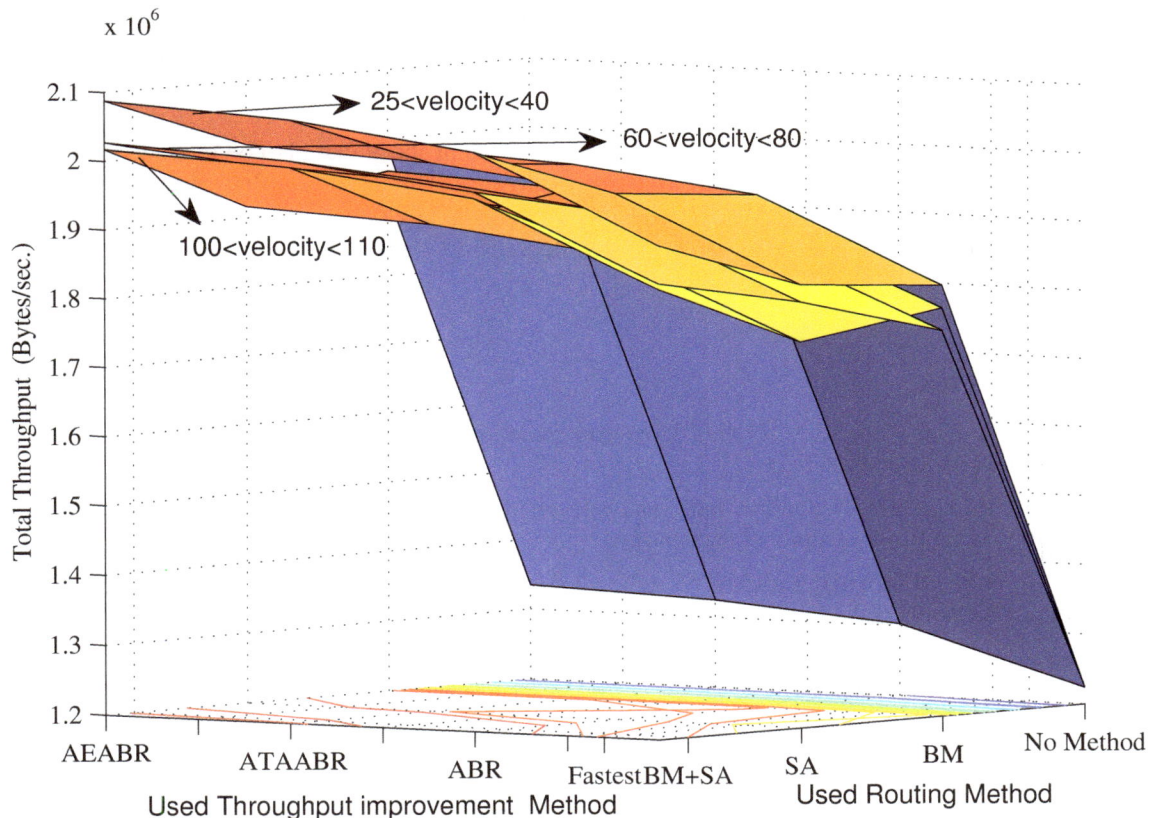

Fig. 7. The throughput improvement by using different; throughput improvement methods vs. routing methods vs. vehicular velocities.

It is shown in Fig.7 that the novel proposed throughput improvement algorithms (BM, SA, BM+SA) are succeeded in improving the system throughput for Fastest Path, Ant-Colony, ABR and other novel Associativity Based Routing algorithms (ATAABR and AEABR).

5. CONCLUSION

To the best of our knowledge, this is the first analytically confirmed event-driven simulative work that investigates the effects of novel routing algorithms on each throughput improvement method for different velocities.

In this work, it is illustrated that the known routing algorithms (ABR, Fastest Path and Ant Colony) and our novel proposed Associativity Based Routing algorithms (AEABR [19] and ATAABR [20]) which generate routes with longer life times, all provide throughput improvement when they are used with the novel OFDMCAF method.

It is shown in this work that, the Fastest Path routing algorithm improves the throughput always better than the Ant Colony routing algorithm. On the other hand all Associativity Based Routing algorithms provide better system throughput performance than others, where proposed AEABR algorithm provides the best performance for all throughput improvement methods and with all vehicular velocities. In case of using no throughput improvement method ABR provides 49% throughput improvement, ATAABR provides 51% throughput improvement, AEABR provides 56% throughput improvement with respect to the Fastest Path algorithm for the vehicular velocities up to 60 km/h and the throughput improvement effect decreases by increased vehicular velocity.

It must be noted that, the results of the methods that we propose can also be evaluated using any other popular event-driven network simulator in the area. The methods, algorithms, formulations and results of this work can be used in designing or analyzing a unicast mobile multimedia network considering the routing algorithms decreasing packet loss rate and improving the throughput.

REFERENCES

[1] Mitola, III J. & Maguire, G, (1999) "Cognitive radio : making software radios more personal", IEEE personal communications,. 13–18.

[2] Mitola, III J. (2000) "Cognitive radio and integrated agent architecture for software defined radio", Dissertation KTH Royal institute of technology.

[3] Hu, X. & Li, B. (2010) "Efficient resource allocation with flexible channel cooperation in OFDMA cognitive radio networks" . INFOCOM, 1–9.

[4] Sirinivasa, S. & Jafar, A. (2006) "The throghput potantial of cognitice radio : A theoretical perspective", ACSSC 2006. 221–225.

[5] Genca, V. & Murphy, S. (2008) "Performance analysis of transparent relays in 802.16j MMR networks". WIOPT 2008. 273–281.

[6] Salameh, H. B. & Krunz, M. (2009) "Spectrum sharing with adaptive power management for throughput enhancement in dynamic access networks." Tech. Rep. TR-UA-ECE-2009-. University of Arizona, Department of ECE.

[7] Kumar, A. (2008), "Mobile broadcasting with WIMAX", Elsevier Inc.

[8] WIMAX FORUM, (2006) "Mobile WIMAX - part 2:a comperative analysis"

[9] WIMAX FORUM, (2006) "Mobile wimax - part 1:a technical overview and performance evaluation."

[10] Preveze, B. & Safak, A. (2010) "Throughput maximization of different signal shapes working on 802.16e mobile multi-hop network using novel cognitive methods." Recent trends in wireless and mobile networks, WIMO 2010 Turkey, 2010. 71–86.

[11] Preveze, B. & Safak, A. (2010), "Throughput improvement of mobile multi-hop wireless networks." International journal of wireless and mobile networks , Vol. 2 No.3 120–140.

[12] Girici, T. (2009) "Asymptotic throughput analysis of multicast transmission schemes", International journal of electronics and communications (AEU) Vol. 63 901–905.

[13] Song, G. & Li, Y. (2006) "Asymptotic throughput analysis of channelaware scheduling". IEEE transactions on communications 54- 10, 1827–1834.

[14] Mach, P. & Bestak, R. (2008), "WIMAX throughput evaluation of conventional relaying", Telecommunication systems Journal Vol. 38 No. 1-2, 11–17.

[15] Deborah, E., Daniel, Z, Li, T. , Yakov R. & Kannan, V. (1995), "Source demand routing packet format and forwarding specification" Mobi Com 1995. 1–7.

[16] Iannone, L. & Fdida, S. (2006), "Can multi rate radios reduce end to end delay in mesh network? a simulation case study", MobiHoc 2006. 15–22.

[17] Yi, X. & Wanye, W. (2008), Finding the fastest path in wireless networks. IEEE ICC 2008. 3188–3192.

[18] Sivajothi, M. & Naganathan, E. N. (2008), "An ant-colony based routing protocol to support multimedia communication in ad-hoc wireless networks", International journal of computer science and network security Vol. 8 No.7, 21–28.

[19] Preveze, B. & Safak, A. (2009), "Alternative Enhancement of Associatively Based Routing (AEABR)for mobile networks. MONAMI 2009. 67–77.

[20] Preveze, B. & Safak, A. (2009) "Associativity Tick Averaged Associativity Based Routing (ATAABR) for real time mobile networks". ELECO 2009.

4

IMPACT OF SELFISH NODE CONCENTRATION IN MANETs

Shailender Gupta[1], C. K. Nagpal[2] and Charu Singla[3]

[1]Department of Electronics Engineering, YMCA University, Faridabad, India
shailender81@gmail.com[1]
[2]Department of Computer Engineering, YMCA University, Faridabad, India
nagpalckumar@rediffmail.com
[3]Department of Electronics Engineering, NGF College of Engg. & Technology, Palwal,
Charu.singhla@gmail.com

ABSTRACT

The communication in Mobile Ad hoc Network (MANET) is multi-hop in nature wherein each node relays data packets of other nodes thereby spending its resources such as battery power, CPU time and memory. In an ideal environment, each node in MANET is supposed to perform this community service truthfully. However this is not the case and existence of selfish nodes is a very common feature in MANETs. A selfish node is one that tries to utilize the network resources for its own profit but is reluctant to spend its own for others. If such behaviour prevails among large number of the nodes in the network, it may eventually lead to disruption of network. This paper studies the impact of selfish nodes concentration on the quality of service in MANETs.

KEYWORDS

Ad hoc Network, Selfish Nodes, Quality of Service

1. INTRODUCTION

Mobile Ad hoc Networks don't rely on extraneous fixed infrastructure and can be installed without base station and dedicated routers. This makes them ideal candidate for rescue and emergency operations [1] and other short term networks. The nodes in these networks have limited battery power and bandwidth, and each node needs the assistance of others to get its packets forwarded. The conventional protocols in MANETs such as WRP [2], DSDV [3], AODV [4] and DSR [5] assume that all the nodes are cooperative and whenever a node receives a request to relay traffic, it always does so truthfully.

However the experience has shown [6, 7, 8, 9] that as the time passes there is a tendency in the nodes in an ad hoc network to become selfish. The selfish nodes are not malicious but are reluctant to spend their resources such as CPU time, memory and battery power for others. The problem is especially critical when with the passage of time the nodes have little residual power and want to conserve it for their own purpose. Thus in MANET environment there is a strong motivation for a node to become selfish.

Marti et. al [7] have defined the characteristics of selfish nodes as follows:

- *Do not participate in routing process*: A selfish node drops routing messages or it may modify the Route Request and Reply packets by changing TTL value to smallest possible value.

- *Do not reply or send hello messages*: A selfish node may not respond to hello messages, hence other nodes may not be able to detect its presence when they need it.

- *Intentionally delay the RREQ packet*: A selfish node may delay the RREQ packet up to the maximum upper limit time. It will certainly avoid itself from routing paths.

- *Dropping of data packet*: A selfish nodes may participate in routing messages but may not relay data packets

The major reason for such behaviour is low residual battery power [6, 8, 9]. It may here be clarified that a selfish node is not malicious and doesn't intend to involve itself in the network damaging activities such as content alteration, spoofing etc. It normally restrains itself from the activities of the other nodes which do not bring any benefit to it.

The literature provides [10, 11, 12, 13, 14] various strategies to deal with such behaviour. These strategies may categorise into two basic categories: Motivation/ incentive based approach and detect and exclude.

The motivation or incentive approach tries to motivate the users of the ad hoc network to actively participate in the forwarding activities. Such a system involves certain amount of money transfer to the relay nodes, on behalf of source or destination, to motivate them to forward messages [11, 15, 16]. One of the motivation/ incentive based approach [8] is based on a virtual currency called nuglet. Every network node has an initial stock of nuglets. Either the source or the destination of each traffic connection use nuglets to pay the relay nodes for forwarding the traffic. The cost of a packet may depend on several parameters such as required total transmission power and the battery status of the intermediate nodes. Packets sent by or destined to nodes that do not have a sufficient amount of nuglets are discarded. The major drawback of this approach is the demand for trusted hardware to secure and maintain the record of the currency at central level. One such protocol, the ad hoc VCG [17] is based on the monetary transfer and discovers an energy-efficient path between the source and the destination. However, the number of messages that must be exchanged in order to find the route to the destination is quite high – in the order of $O(n^3)$, where n is the number of network nodes.

Detect and exclude strategy avoids selfish nodes from the routing paths. This scheme uses two types of trust namely first hand trust and second hand trust.

- First hand trust: The node's personal observation about the neighbouring nodes.

- Second hand trust: The observation communicated by neighbouring node about the other neighbours of the network.

Watchdog and Pathrater [7] is a mechanism based on detect and exclude principle to deal with the selfish nodes. It uses Dynamic Source Routing [18] as base protocol. It has two components: Watchdog and Pathrater. The Watchdog is responsible for detecting selfish nodes that do not forward packets. The Pathrater assigns different rating to the nodes based upon the feedback that it receives from the Watchdog. Each node in the network buffers every transmitted packet for some time. During this interval, the node places its wireless interface into the promiscuous mode in order to overhear whether the next node has forwarded the packet or not and ratings are developed. These ratings are then used to select routes consisting of nodes with the highest forwarding rate.

SORI [22] is a protocol based on detect and exclude mechanism. It makes two record, the local evaluation record (First hand trust) and the overall evaluation record based on the reputation index given by the nodes about their neighbours. Each node in the network maintains tables of first and second hand trust of their neighbouring nodes. Based on these tables the trust of a node is calculated and then action is taken against the selfish nodes

CONFIDANT [19] protocol adds trust manager and reputation index to the Watchdog and Pathrater mechanism. Each node in the network maintains two lists to deal with the selfish nodes. The nodes which behave rationally are kept in the friends list and the nodes which drop the packets or tamper them are kept in the black list. These lists are exchanged by the neighbouring nodes. Based on these list trust of a particular node is calculated. Whenever the trust value for a particular node falls below a certain threshold the protocol stops forwarding packets of that node.

Y. Zhang et. al. [13, 20] designed a intrusion detection system (IDS) for MANETs that consists of Local components: data collection, detection and response and Global components: cooperative detection and global response.

Collaborative Reputation Mechanism (CORE) [21] is similar to the distributed IDS by Zhang et al. and consists of local observations that are combined and distributed to calculate a reputation value for each node. Based on this reputation, nodes are allowed to participate in the network or are excluded. In their work, the authors specify in detail how the different nodes should cooperate to combine the local reputation values to a global reputation and how they should react to negative reputations of nodes. An aspect that is not clearly stated in the work of Y. Zhang et. al.

2. THE CONCEPT AND THE ASSOCIATED WORK

All these strategies are generic and static do not take into consideration the concentration level of selfish nodes into network which may change dynamically. We are of the view that the strategy to deal with this selfish behaviour should be dependent on their concentration level in the network as the impact they have on the network activity will be different at different level of their concentration.

This paper studies the impact of selfish nodes concentration [0-100%] on the various Quality of Service (QoS) parameters [23, 24, 25, 26] related to performance of ad hoc network using MATLAB simulation. The QoS parameters taken into consideration are as follows:

- *Throughput*: Percentage of packets received by the destination to the number of packets sent by the source.

- *Hop count*: Defined as the number of intermediate hops from source to destination.

- *Packet dropped*: Measure of the number of packets dropped by the routers due to various reasons.

- *Probability of Reachability*: Fraction of possible reachable routes to the all possible routes between all different sources to all different destinations.

The purpose of the proposed work is to help the various protocol designers to enable them in incorporating the dynamic strategies to deal with selfish nodes as their concentration varies.

3. SIMULATION AND RESULTS

A simulator was designed in MATLAB in which an area of 30 X 30 sq. unit's size was chosen. In total 40 nodes were distributed randomly in the given area, using *randint* function that uniformly distributes the random numbers, as shown in Figure 1. To create the route between a pair of random source and destination *Dijkstra's shortest path algorithm* was used. To avoid the selfish nodes in the path all the selfish nodes were made unreachable during that particular iteration. Figure 1. shows the snapshot of the simulation process wherein the red lines shows the shortest path when no node is selfish and the green lines shows the shortest path after the avoidance of selfish nodes. At low concentration of selfish nodes it is more likely green and red lines will be same and as the concentration increases they are likely to be different. Thus at 0% concentration of selfish nodes the route found between a pair of source and destination is shortest and at k% concentration the route found is shortest as if no selfish nodes were present. Thus with the increase in concentration of selfish nodes

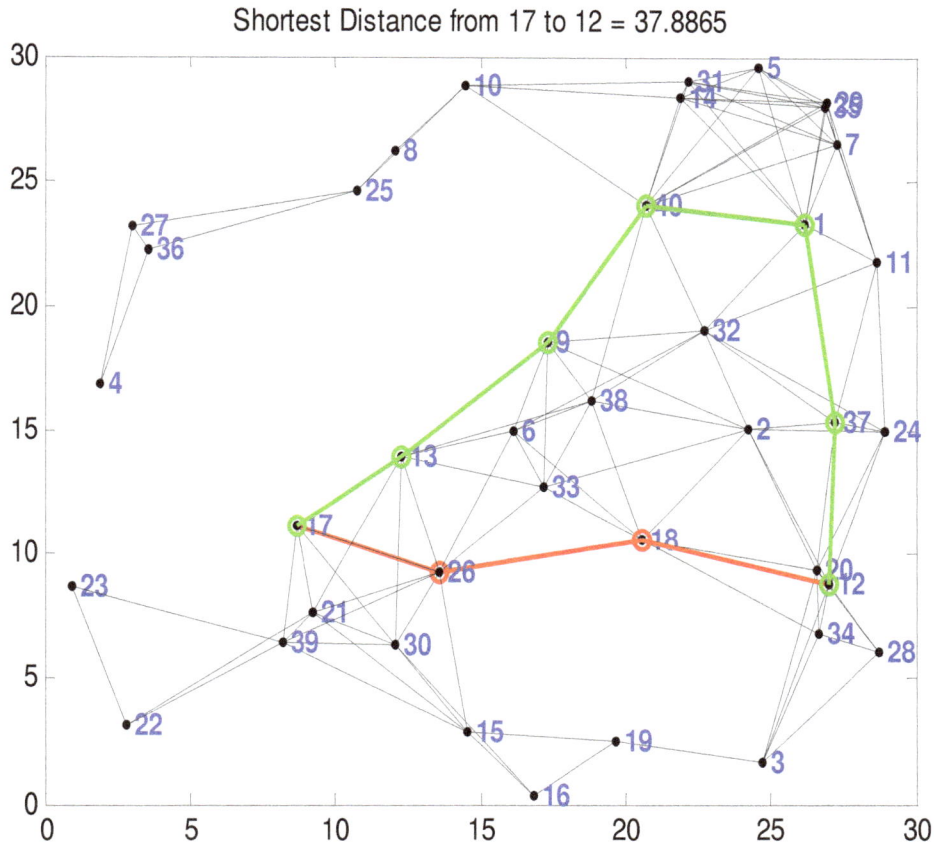

Figure 1. Ad hoc Network

- The average hop count is likely to increase

- The packet drop rate is likely to increase

- The average throughput is likely to decrease

- The probability of reachability is likely to decrease

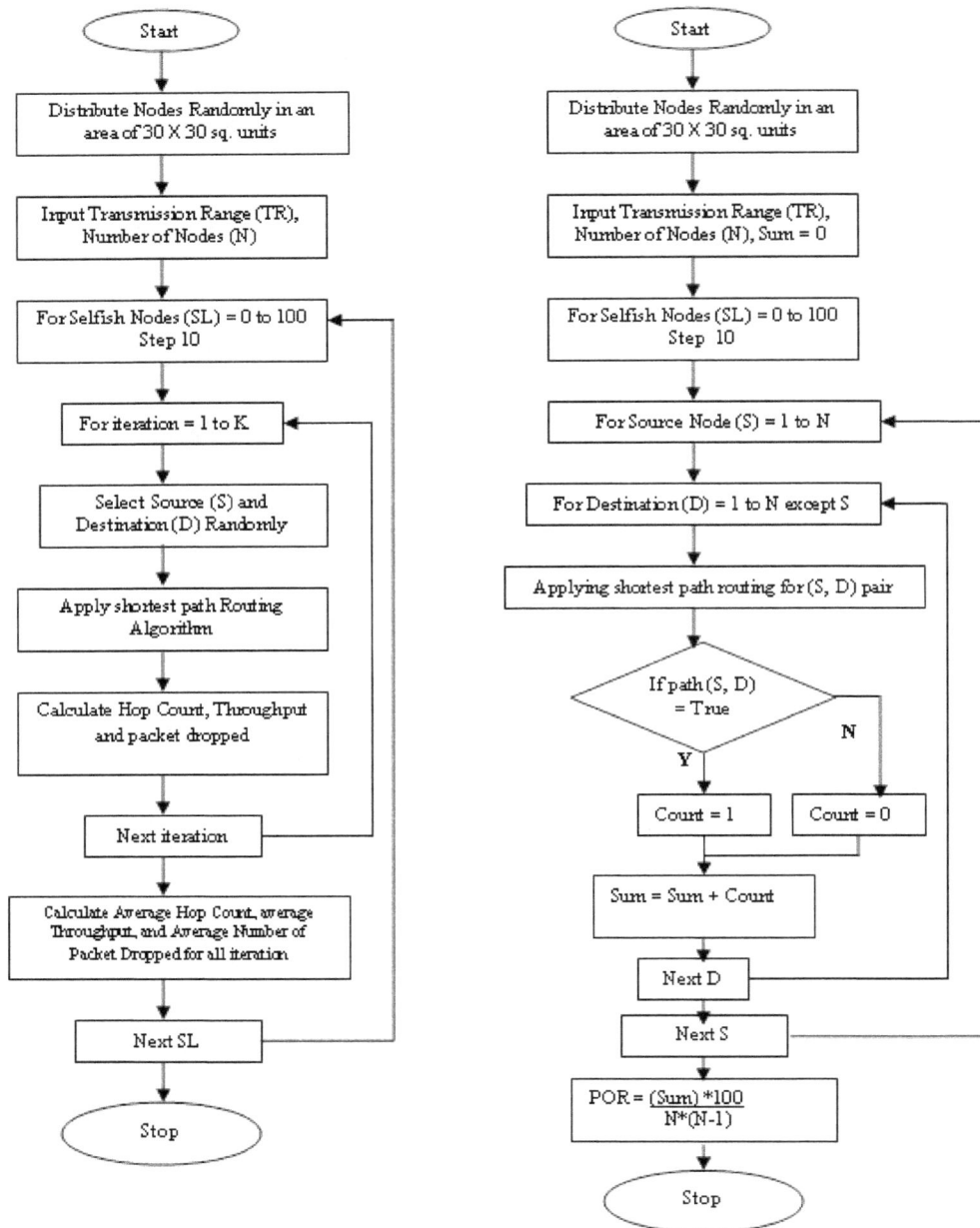

Figure 2. Flowchart of Simulators

Keeping these aspects in view and our simulator tried to make a quantitative assessment of the impact of selfish nodes on the above mentioned QoS parameters. This assessment will enable the designer of the protocols to combat the selfish node problem in the effective manner in an environment where the concentration of selfish node is changing dynamically. The quantitative assessment will enable the designer to make rational decisions while designing the protocol.

Figure 2. shows the flow chart for implementing the simulator on MATLAB.

3.1. IMPACT ON AVERAGE HOP COUNT:

Figure 3. shows the impact of increase in average hop count as the concentration of selfish nodes increases. The average hop count is almost same when the selfish nodes are up to 10% of the total number of nodes. The average hop count increases to 2.5 times when the selfish nodes are nearly 100%. If the route formation does not occour, then in that the maximum hop count (16) was taken.

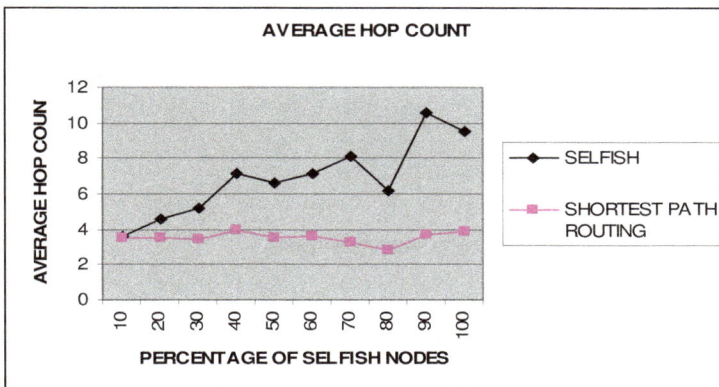

Figure 3. Impact of Selfish Nodes on Average Hopcount

3.2. IMPACT ON NUMBER OF PACKETS DROPPED

Figure 4. shows the impact of selfish nodes concentration on the percentage of packets dropped. There is no remarkable change in the percentage packet dropped when the selfish node concentration is up to 10%. It reaches to a maximum value of nearly 60% when the selfish node concentration is nearly 90%.

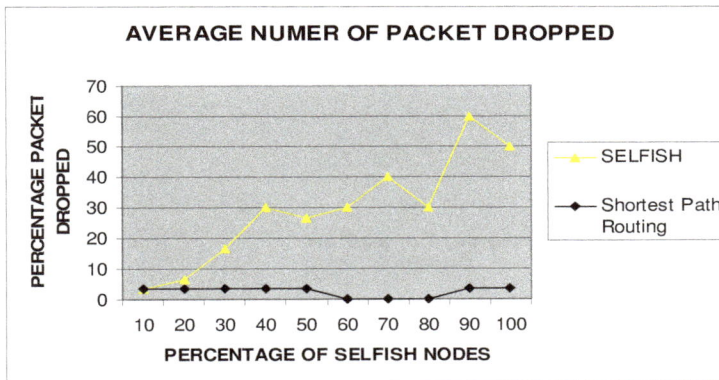

Figure 4. Impact of Selfish Nodes on Average Packet Dropped

3.3. IMPACT ON THROUGHPUT

The results shows that there is no significant decrease in average throughput when the number of selfish nodes up to 10% of the total number of nodes. There is a generic trend of decrease in average throughput with the increase in concentration of selfish nodes. Since the process of source destination pair selection is random there are some fluctuations in the results as shown in Figure 5. It can be seen that the average throughput doesn't fall to zero even if all the nodes are selfish. The reason for this is that even at 100% selfish behaviour the communication between two immediate neighbours, as source and destination, still survives.

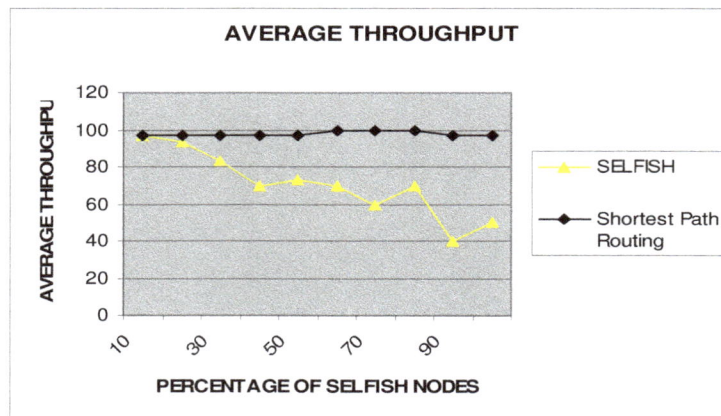

AVERAGE THROUGHPUT

Figure 5. Impact of Selfish Nodes on Average Throughput

3.4. IMPACT ON PROBABILITY OF REACHABILITY

As expected the probability of reachability is nearly 100% when no node is selfish. As the number of selfish nodes increases the probability of reachability decreases but never reaches zero because even at 100% selfish behavior the communication between two immediate neighbors, as source and destination, still survives. Figure 6. shows the impact of selfish node concentration on probability of reachability.

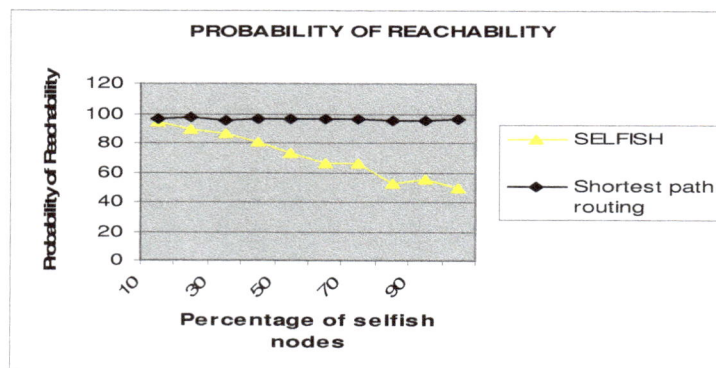

PROBABILITY OF REACHABILITY

Figure 6. Impact of Selfish Nodes on Probability of Reachability

4. CONCLUSIONS

The problem of selfish nodes is very common in ad hoc networks. The major reason for the selfishness is the loss of power with time. As the time passes away the nodes loose their battery power and in a disaster hit area or battle field area recharging may not be easily possible. The experimental results show that up to a concentration level of 10%, selfish nodes do not have remarkable negative effect on the network activities. As the concentration increases QoS becomes poorer and poorer though the network never comes to halt even if the selfish node concentration reaches to nearly 100%. The average hop count reaches to a maximum 2.5 times, Probability of Reachability and throughput comes down to nearly 50% at its peek and percentage of packet drop goes up to nearly 60% at the most. Since the concentration of selfish nodes increases with time, the routing protocols can be designed in such a manner that they can react to dynamic change in the concentration of selfish nodes.

ACKNOWLEDGEMENTS

The authors would like to thank everyone, just everyone!

REFERENCES

[1]. Brian B. Luu, Barry J. O'Brien, David G. Baran, and Rommie L. Hardy, "A Soldier-Robot Ad Hoc Network" Proceedings of the Fifth Annual IEEE International Conference on Pervasive Computing and Communications Workshops (PerComW'07).

[2]. Murthy, S. and J.J. Garcia-Luna-Aceves, "An Efficient Routing Protocol for Wireless Networks", ACM Mobile Networks and App. J., Special Issue on Routing in Mobile Communication Networks, Oct. 1996, pp. 183-97.

[3]. C. E. Perkins and P. Bhagwat, "Highly dynamic Destination-Sequenced Distance-Vector Routing (DSDV) for mobile computers," ACM Computer Communication Review, Vol. 24, No.4, (ACM SIGCOMM'94) Oct. 1994, pp.234-244.

[4]. C.E. Perkins and E.M. Royer, "Ad hoc on demand Distance Vector routing," mobile computing systems and applications, 1999. Proceedings. WMCSA '99. Second IEEE Workshop on, 1999, p90 - p100.

[5]. D. Johnson, D. A. Maltz, "Dynamic source routing in ad hoc wireless networks," in Mobile Computing (T. Imielinski and H. Korth, eds.), Kluwer Acad. Publ., 1996.

[6]. L. Blazevic, L. Buttyan, S. Capkun, S. Giordano, J. P. Hubaux, J. Y. Le Boudec, "Self-Organization in Mobile Ad-Hoc Networks: the Approach of Terminodes," IEEE Communications Magazine, Vol. 39, No. 6, June 2001.

[7]. S. Marti, T. J. Giuli, K. Lai, M. Baker, "Mitigating Routing Misbehavior in Mobile Ad Hoc Networks," Proc. of MobiCom 2000, Boston, August 2000.

[8]. L. Buttyán, J.-P. Hubaux, "Nuglets: a Virtual Currency to Stimulate Cooperation in Self-Organized Mobile Ad Hoc Networks," Technical report No. DSC/2001/001, Swiss Federal Institution of Technology, Lausanne, January 2001. http://icawww.epfl.ch/hubaux/.

[9]. L. Buttyán, J.-P. Hubaux, "Stimulating Cooperation in Self-Organizing Mobile Ad Hoc Networks," Technical Report No. DSC/2001/046, Swiss Federal Institution of Technology, Lausanne, 31 August 2001. http://icawww.epfl.ch/hubaux/.

[10]. Levente Butty´an and Jean-Pierre Hubaux. Nuglets: a Virtual Currency to Stimulate Cooperation in Self-Organized Mobile Ad Hoc Networks. Technical Report DSC/2001/001, EPFL-DI-ICA, January 2001.

[11]. Levente Butty´an and Jean-Pierre Hubaux. Stimulating Cooperation in Self-Organizing Mobile Ad Hoc Networks. ACM/Kluwer Mobile Networks and Applications, 8(5), October 2003.

[12]. Sheng Zhong, Jiang Chen, and Yang Richard Yang. Sprite: A simple, cheat-proof, credit-based system for mobile ad-hoc networks. In Proceedings of IEEE Infocom '03, San Francisco, CA, April 2003.

[13]. Yongguang Zhang and Wenke Lee. Intrusion detection in wireless ad-hoc networks. In Mobile Computing and Networking, pages 275–283, 2000. also available as http://citeseer.nj.nec.com/zhang00intrusion.html.

[14]. Yongguang Zhang, Wenke Lee, and Yi-An Huang. Intrusion Detection Techniques for Mobile Wireless Networks. to appear in ACM Wireless Networks (WINET), 9, 2003. also available as http://www.wins.hrl.com/people/ygz/papers/winet03. pdf.

[15]. L. Buttyan and J. Hubaux, "Enforcing Service Availability in Mobile Ad-Hoc WANs", in Proceedings of IEEE/ACM Workshop on Mobile Ad Hoc Networking and Computing (MobiHOC), Boston, August 2000.

[16].   N. Ben Salem, L. Buttyan, J.P. Hubaux, M. Jakobsson, "A Charging and Rewarding Scheme for Packet Forwarding in Multi-hop Cellular Networks", in Proc. ACM MobiHoc 03, pp. 13–24, 2003.

[17].   L. Anderegg, S. Eidenbenz, "Ad hoc-VCG: A Truthful and Cost-Efficient Routing Protocol for Mobile Ad hoc Networks with Selfish Agents", Proc. ACM Mobicom, pp. 245–259, 2003.

[18].   D.B. Johnson, D.A. Maltz, J. Broch, et al. DSR: The dynamic source routing protocol for multi-hop wireless ad hoc networks. Ad hoc networking, 5:139–172, 2001.

[19].   Buchegger S, Le Boudec JY. Performance analysis of the CONFIDANT protocol. In Proceedings of 3rd ACM International Symposium, on Mobile Ad Hoc Networking and Computing, June 2002.

[20].   Yongguang Zhang, Wenke Lee, and Yi-An Huang. Intrusion Detection Techniques for Mobile Wireless Networks. to appear in ACM Wireless Networks (WINET), 9, 2003. also available as http://www.wins.hrl.com/people/ygz/papers/winet03. pdf.

[21].   Pietro Michiardi and Refik Molva. Prevention of Denial of Service attacks and Selfishness in Mobile Ad Hoc Networks. http://www.eurecom.fr/ michiard/pub/michiardi adhoc dos.ps.

[22].   He Q, Wu D, Khosla P. SORI: A secure and objective reputation- based incentive scheme for ad-hoc networks. In Proceedings of IEEE WCNC2004, March 2004.

[24].   C. Zhu and M. S. Corson, "QoS routing for mobile adhoc networks", IEEE INFOCOM'02.

[25].   C.R. Lin and J.S. Liu, "QoS routing in ad hoc wireless networks", IEEE Journal on Selected Areas in Communications, vol 17, no. 8, August 1999, pp. 1426-1438.

[26].   S. Chen and Klara Nahrstedt, "Distributed quality-ofservice routing in ad hoc networks", IEEE Journal on Selected Areas in Communications, vol.17, no. 8, August 1999, pp.1488-1505.

[27].   Shin Yokoyama, Yoshikazu Nakane, Osamu Takahashi and Eiichi Miyamoto "Evaluation of the Impact of Selfish Nodes in Ad Hoc Networks and Detection and Countermeasure Methods," Proceedings of the 7th International Conference on Mobile Data Management (MDM'06) IEEE 2006.

5

WEB BASED EMBEDDED ROBOT FOR SAFETY AND SECURITY APPLICATIONS USING ZIGBEE

V. Ramya[1] and B. Palaniappan[2]

[1]Assistant Professor Department of Computer Science and Engineering,
Annamalai University, Chidambaram, Tamilnadu
E-Mail: ramyshri@yahoo.com

[2]Dean, FEAT, Head Department of Computer Science and Engineering,
Annamalai University, Chidambaram, Tamilnadu
E-Mail: bpau2002@yahoo.co.in

Abstract

This project proposed an embedded system for safety and security purpose robot using zigbee communication and web server. The robot has sensors for detecting Gas leakage and intruder detection. MQ6 Gas sensor detects the presence of bio hazardous gases like LPG, iso-butane, propane, LNG and alcohol, and the PIR sensor detects only the living organism (Intruder). The sensor details are first sent to the microcontroller which resides at the robotic side and then sent to the local system through Zigbee. The system also provides an audio and visual alarm to alert about the critical situation for the safety and security purpose. This robot also has a battery powered wireless AV camera which provides robotic in front environment information to the Local and remote system and performs the audio and video streaming through web server. The robotic movement is controlled remotely from the local system by using the front end application VB 6.0. The Zigbee (IEEE 802.15.4) supports a frequency range of 2.4GHZ, 9600 baud rate with 256Kb of flash memory. It supports the range of 400m in open-air, line-of-sight, outdoor environment. This proposed system is used wherever people cannot go or where things doing too dangerous for humans to do safely. That is the robot can move and reach to the high destiny gas leakage region.

Key words: Audio-Video streaming, Intruder, Remote system, Robot, PIR Sensor, Web Server, Zigbee.

1. INTRODUCTION

Mini robot is an autonomous security robot. The robot's design specifications may vary according to the given application. An embedded system is designed to perform specific control functions within a larger system, often with temporal constraints. It is embedded as part of a complete device often including hardware and mechanical parts. Embedded systems contain processing cores such microprocessors, microcontrollers and discrete processors. The key characteristic, however, is being dedicated to handle a particular task. Since the embedded system is dedicated to specific tasks, design engineers can optimize it to reduce the size and cost of the product and increase their reliability and performance [1].

1.1 Proposed system

The proposed robot is easy to design and implement both in hardware and software aspects. It uses low cost microcontroller, high sensitivity gas and PIR sensors, wireless AV camera and zigbee to support reliable and robust wireless communication network. But in existing system they have used high cost IP camera. The AT89C51 microcontroller is embedded with embedded C program which processes the received sensor data and provides safety and security alarm through zigbee communication. The mobile robot is a battery powered and controlled remotely through zigbee. At the local system, the front end is designed using VB6 which is simple coding and easy to understand. The existing system does not support live AV streaming but our system provides live AV streaming to the remote web server through tin cam. The robot integrates both safety and security functions and is useful in variety of applications like industries, resorts, government and non government organizations. This intelligent robot is mainly useful in rescue operations, which detects the alive human in disaster situations and in war fields, and also used in intelligent security purposes.

2. ROBOTIC SYSTEM DESIGN PROCESS

Figure 1 shows the embedded system design process and has five major levels of abstraction. At each and every level there are three tasks like analyze, refine and verify were performed to ensure the system requirements and specifications. There are two ways in the design methodology and are Top-down and Bottom-up method [1, 13] and this work adopts the Top-down method.

2.1 Requirements

An informal description from the customers was gathered and is known as requirements then the requirements are refined into a specification that contains enough information to begin the system architecture. Requirements may be functional or nonfunctional. We must of course capture the basic functions of the embedded system, but functional description is often not sufficient. Typical nonfunctional requirements include: performance, cost, power and physical size and weight and these requirements are given in the requirement form shown in the Table 1.

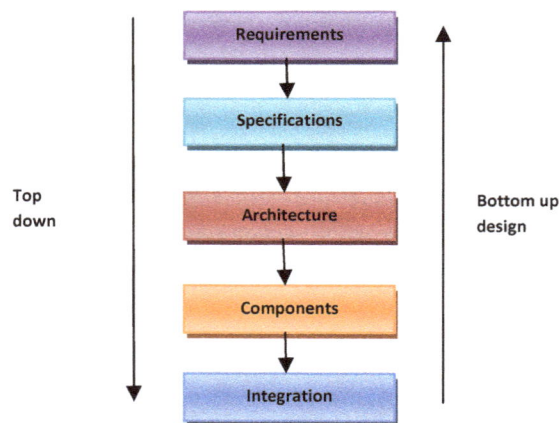

Figure 1: Major levels of abstraction in the design process

Functional Requirements:
- Sensoring requirements like the Gas, Human Detection and Audio-Video Streams.
- Alarm Monitoring requirements [If GAS, Human are detected].

- Signal conditioning requirements.
- Movement through Relay for directions that switches ON/OFF.

Table 1: Requirement form

Requirements	Descriptions
Name	Embedded Robot for Safety and security applications using Zigbee.
Purpose	ROBOT monitors the environment; provide Audio- video streaming at remote system. Raises alarm during critical situations for the safety and security purpose.
Inputs	LPG Gas Sensor, PIR Sensor, Audio and video signal.
Outputs	Remote PC display, Relays, LEDs
Functions	Depending on the Gas and PIR sensor provide alert message to local system and switch ON/OFF the relays for Robotic wheel movement. Perform Audio-Video streaming both at local and remote system.
Performance	Updates sensor data to the local system at every 5 to 8 seconds and update AV information to the local and remote system at every 0.5seconds.
Manufacturing Cost	Approx. 125USD
Power	12v, 9Amps
Physical size and Weight	11"X6.5" and 1.700kgs

2.2 Specification

The Specification says only what the system does and does not specify how to implement. In this proposed work the specification of the system includes,

- Data received from Zigbee (100M)
- User interface
- Sensor data to the Microcontroller
- PC Display
- Alarm Monitoring
- Audio and Video from wireless AV camera

2.3 Architecture

The Architecture describes how the system implements the functions which are specified in the specification level. The system architecture is further refined in to Hardware and software architecture which describes the components we need to build the entire system. Architectural descriptions must be designed to satisfy both functional and non-functional requirements. Figure 2 shows the hardware module in which the sensors send the signals to the local system through Zigbee, the local system process the data and send the control signals to the Robot and switches the relay according to the conditions for ON or OFF the Motors [2]. The Zigbee connected to the microcontroller which sends the data from the ROBOT to the local system and the Zigbee at the local system receives the data and displays the output. Fig 3 shows software architecture of the robot.

2.4 Components

The Components in general include both designing the hardware and software components. Some of the components will be ready made, for example CPU, memory and I/O. First we have to decide that either to buy the components which are readymade or to build by ourselves. If we buy the components then the design time will be reduced and also increases the implementation speed. The components used in this work are discussed in section 3.The basic components involved in this project are,

- Zigbee
- Micro controller
- PIR Sensor
- LPG Gas Sensor
- Relays
- Wireless AV Camera
- DC Motors

2.5 Integration

The System Integration is not simply plugging everything together but also finding the bug at this stage. While testing the system, it is difficult to find why things are not working properly and hence it is hard to find and fix the bug. Due to limited facility at the target system, we have to go to host system each and every time for testing. As for as the embedded system concern the system integration is a challenging task.

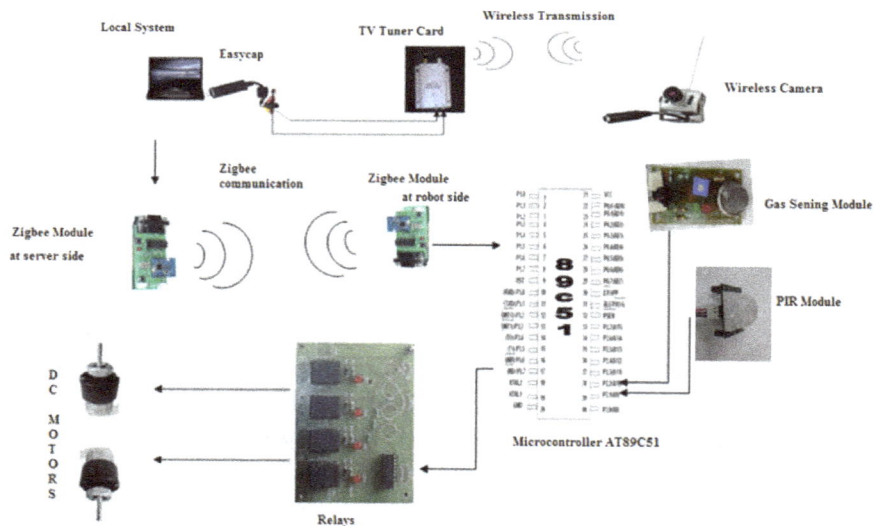

Figure 2: Hardware Architecture

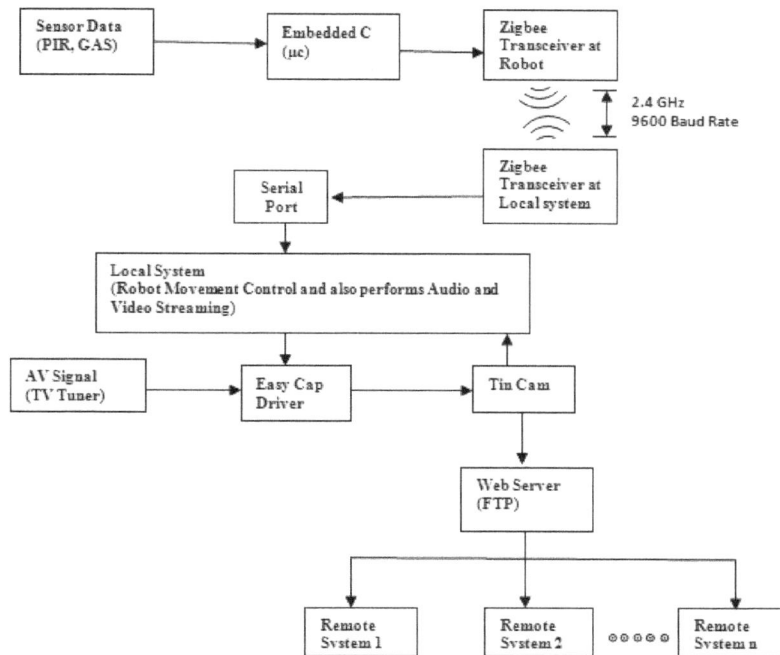

Figure 3: Software Architecture

3. HARDWARE DESCRIPTIONS

3.1. Microcontroller

AT89C51 microcontroller is used in this work which is an 8-bit microcontroller and has 4KB of Flash memory and 128 bytes of RAM. The on-chip Flash allows the program memory to be reprogrammed in-system or by a conventional nonvolatile memory programmer. The Atmel AT89C51 is a powerful microcomputer which provides a highly-flexible and cost-effective solution to many embedded control applications. AT89C51 has four ports designated as P_1, P_2, P_3 and P_0 and all these ports are 8-bit bi-directional ports; they can be used as both input and output ports. Whenever the port P0 and P2 connected to an external memory they provide both low byte and high byte addresses, respectively. Port 3 has multiplexed pins which can be used for serial communication functions, hardware interrupts, timer inputs and read/write operation from external memory. It has a total of six interrupts. In this proposed system the PIR sensor module is connected with the port $P_{0.1}$/AD0 at pin number 39, Gas sensor module is with port $P_{0.1}$/AD1 at pin number 38, zigbee module's transmitter and receiver is connected with port $P_{3.0}$/RXD at pin number10 and $P_{3.1}$/TXD at pin 11 and relay driver ULN2003A is connected with the ports $P_{1.4}$, $P_{1.5}$, $P_{1.6}$, $P_{1.7}$ of pins 5, 6, 7 and 8 respectively.

3.2. PIR Sensor

A Passive Infra Red sensor (PIR) is a device used to detect motion by receiving infrared radiation. A PIR detector Combined with a Fresnel lens (FL65) is mounted on a compact size PCB together with an analog ICSB0081 and limited components to form the module. A Fresnel lens is a Plano Convex lens that has been collapsed on it to form a flat lens that retains its optical characteristics but is much smaller in thickness and therefore has less absorption losses. The FL65 Fresnel lens is made of an infrared transmitting material that has an IR transmission range of 8 to 14 µm that is most sensitive to human body radiation. The FL65 has a focal length of 0.65 inches from the lens to the sensing element [15]. Due to the high sensitivity of PIR sensor device, the TTL output can be directly connected to the micro controller; here it is connected with the port $P_{0.1}/AD0$ at pin number 39 of the AT89C51 microcontroller. Fig 4 and 5 shows the PIR sensor pin description and the sensor detection range. The PIR sensor detection and communication with the local system is explained in the flowchart which is shown in figure9.

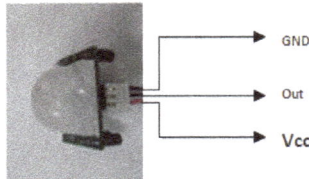

Figure 4: PIR Sensor   Figure 5: PIR Sensor Range

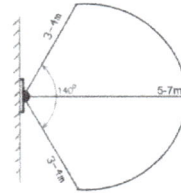

Figure 5 shows The PIR detects the intruder only if he is in the lens coverage range of 140°, which means 5 to 7 meters from the centre of lens.

3.3. MQ6- Gas Sensor

The MQ6 Gas Sensor Module determines the presence of LPG gas and other bio hazardous gases. The proposed system interfaces the gas sensor module with the microcontroller at the port $P_{0.1}/AD1$ at pin number 38 of the AT89C51host microcontroller. The onboard microcontroller provides initial heating interval after power up and then starts to measure LPG sensor output [3]. If the LPG or the other gas level is in the range of 200 to 10000ppm, then it will inform to the host controller by pulling the output Pin to high and starts to blink the onboard status LED. This sensor module is placed in the movable robot which can be used to detect gas leakage at home and industry, and are suitable for detecting of LPG, iso-butane, propane, LNG. A Sensitive tuner is always available in this sensor module, which is mainly used to manually adjust to set the density of the Gas [4]. The Gas sensor detection and communication with the local system is explained in the flowchart which is shown in figure9. The prototype of the gas sensor module used in this project is shown figure 6. Table 2 shows the sensitivity characteristics, its standard and environmental working conditions.

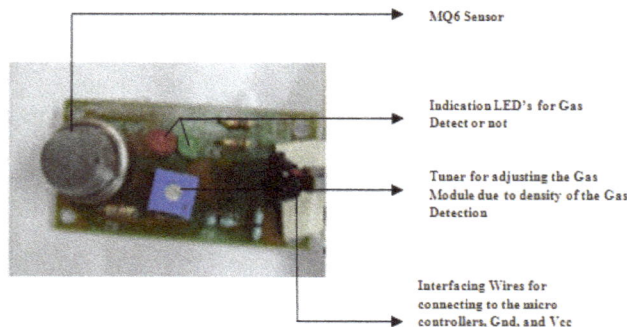

Figure 6: Gas Sensor module

Table 2: MQ6-Gas Sensor Specifications

A. Standard work condition

Symbol	Parameter name	Technical condition	Remarks
Vc	Circuit voltage	5V±0.1	AC OR DC
V_H	Heating voltage	5V±0.1	ACOR DC
P_L	Load resistance	20KΩ	
R_H	Heater resistance	33 Ω ±5%	Room Tem
P_H	Heating consumption	less than 750mw	

B. Environment condition

Symbol	Parameter name	Technical condition	Remarks
Tao	Using Tem	-10°C-50°C	
Tas	Storage Tem	-20°C-70°C	
R_H	Related humidity	less than 95%Rh	
O_2	Oxygen concentration	21%(standard condition)Oxygen concentration can affect sensitivity	minimum value is over 2%

C. Sensitivity characteristic

Symbol	Parameter name	Technical parameter	Remarks
Rs	Sensing Resistance	10KΩ - 60KΩ (1000ppm LPG)	Detecting concentration scope: 200-10000ppm
α (1000ppm/ 4000ppm LPG)	Concentration slope rate	≤0.6	LPG , iso-butane, propane, LNG
Standard detecting condition	Temp: 20°C±2°C Humidity: 65%±5%	Vc:5V±0.1 Vh: 5V±0.1	
Preheat time	Over 24 hour		

3.4. Zigbee Module

The CC2530 zigbee module used in this work is a true system-on-chip solution tailored for IEEE 802.15.4 and is suitable for the low power applications. The CC2530 combines the excellent performance of a leading RF transceiver with an industry-standard enhanced 8051 MCU, in-system programmable flash memory, 8-KB RAM. The CC2530 comes in four different flash versions: CC2530F32/64/128/256, with 32/64/128/256 KB of flash memory, respectively. This project uses 256K flash memory. Short transition times between operating modes further ensures low energy consumption [5]. The range of CC2530 highly depends on antenna design, product enclosure, physical environment including obstructions obstacles environment like temperature and humidity of the air. It is able to achieve 99% packet transmission success with 400m in open-air, line-of-sight, outdoor environment. The Zigbee module is shown in fig 7. The zigbee module's transmitter and receiver is connected with port $P_{3.0}$/RXD at pin number10 and $P_{3.1}$/TXD at pin 11 of the AT89C51 microcontroller. The communication between robot and the local system is a wireless communication using Zigbee. Figure 7(a) shows interfacing RS232 with Zigbee at Local System and figure7 (b) shows interfacing Zigbee at Robot Side.

3.4.1 CC2530 Key Features
- Up to 256 KB Flash/8 KB of RAM
- Excellent link budget (102 dBm)
- 49 dB adjacent channel rejection (best in class)
- Four flexible power modes for reduced power consumption
- Powerful five-channel DMA

Figure 7(a) Interface RS232 with Zigbee at Local System Side

Figure7 (b) Interface Zigbee at Robot

Figure 7: Zigbee module

3.5. DC Motor

A 12V DC geared motors is very easy to use and available in standard size. Nut and threads on shaft is easily connected and internal threaded shaft are easily connecting it to wheel. The 12V DC Geared Motor is used in variety of robotics applications which is available in wide range of RPM and Torque [6]. Fig 8 shows how the DC motor is connected with robotic wheel.

3.5.1 Features

- 30RPM 12V DC motors with Gearbox
- 6mm shaft diameter with internal hole
- 125gm weight
- Same size motor available in various rpm
- 2kgcm torque
- No-load current = 60 mA(Max), Load current = 300 mA(Max)

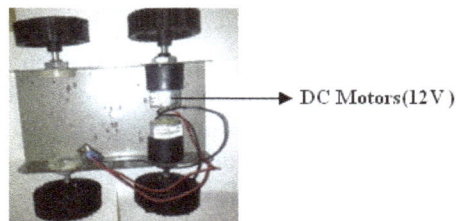

Figure 8: DC motor with Robotic Wheel

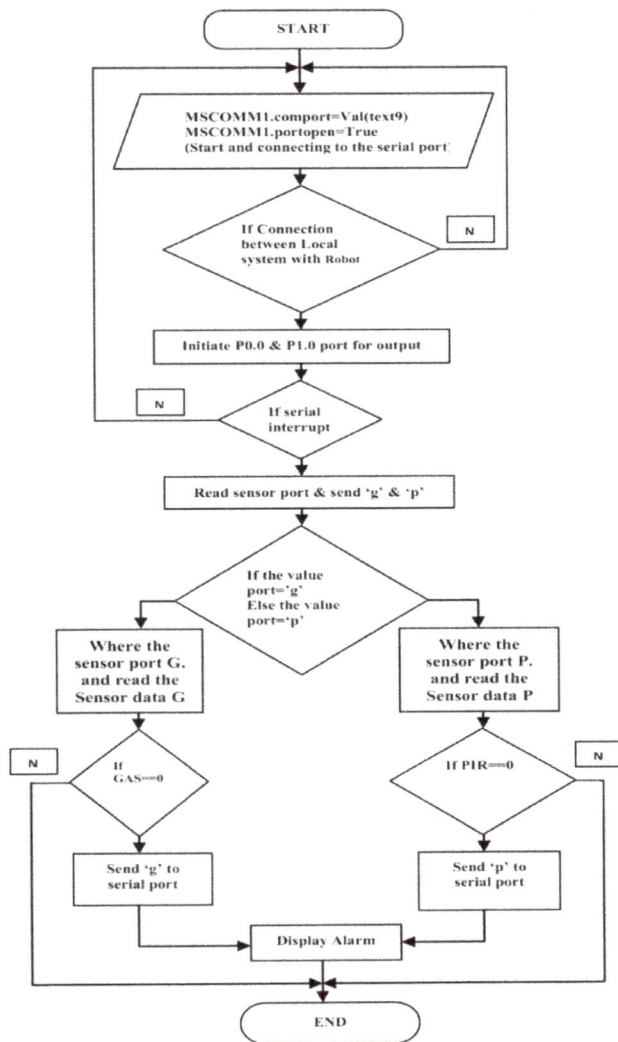

Figure 9: Flow chart for Gas and PIR sensor module

3.6. Relays

A relay is an electrically operated switch. Current flowing through the coil of the relay creates a magnetic field which attracts a lever and changes the switch contacts. Relays are used where it is necessary to control a circuit by a low-power signal. The four relays for controlling the robotic movement is shown in fig 10.

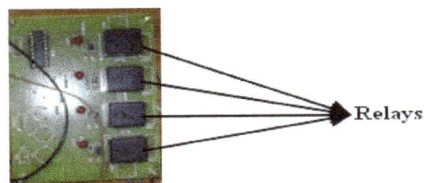

Figure 10: Relays used in the prototype

Table 3: Relay Specification details for Robot Wheel Movement

Relays controlling the Motor directions				
R1	R2	R3	R4	Direction
OFF	OFF	OFF	ON	Left
ON	ON	ON	ON	Forward
ON	OFF	OFF	OFF	Right
OFF	OFF	OFF	OFF	Stop
ON	OFF	OFF	ON	Back

Robot wheel movement is the main process to move the robot wherever the user wants. And the user can easily move the robot by programming the relay control procedure that is fed in to the controller [7]. The relays are driven with the help of ULN2003A driver. It is a driver which is mostly used to activate and deactivate the relays. The following process is maintained by the controller and is accessed by the user. The relays working principle is explained in the flowchart which is shown in figure 11, and also in the table 3. The relay functions are as follows, If relay1=relay2=ON (relay3=relay4=0), the right side motor starts; then the robot turns to **Left** side direction. And relay3=relay4=ON (relay3=relay4=0), the left side motor starts so the robot will be move on **Right** side direction. These both cases are possible only when the other relays value are=0. If relay1= relay4=0 (relay2= relay3=OFF), both the motors starts and the robot will be move **Forward** direction, similarly relay2=relay3=ON (relay1=relay4=OFF), the robot move in **Backward** direction. If all the relays, that is relay1= relay2= relay3= relay4=0 then the motor is said to be in stop condition and hence the robot cannot move. It will remains stop, until anyone of the relay gets started.

3.7. Wireless AV Camera

It is a small and having delicate appearance, good performance with high-quality picture and sound transmitting and receiving. It supports minimum of 100m transmission distance without block and can be used on TV, monitor, LCD, etc. including adaptive bracket and supports easy installation. Wireless video communications are shown in fig 12. The AV signal from the camera is sent to the TV tuner which is connected to the local system through the easy cap (USB 2.0 Grabber). The Audio-Video streaming is done at the local and also at the remote system. This process is explained in the section 6.1 and 6.2.

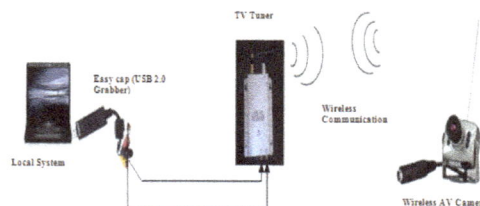

Figure 12: Wireless video communication

3.7.1 Technical Parameters of Transmitting Unit

- Output Power: 50MWOutput Frequency:1.2G/2.4GTransmission
- Signal: Video, Audio Linear Transmission Distance: 50-100M
- Voltage: DC+9V
- Current: 300mA
- Power Dissipation : <=640MW
- Validity pixel: PAL 628x582

- Scan Frequency: PAL/CCIR50Hz
- Sensitivity: +18dB-AGL ON-OFF

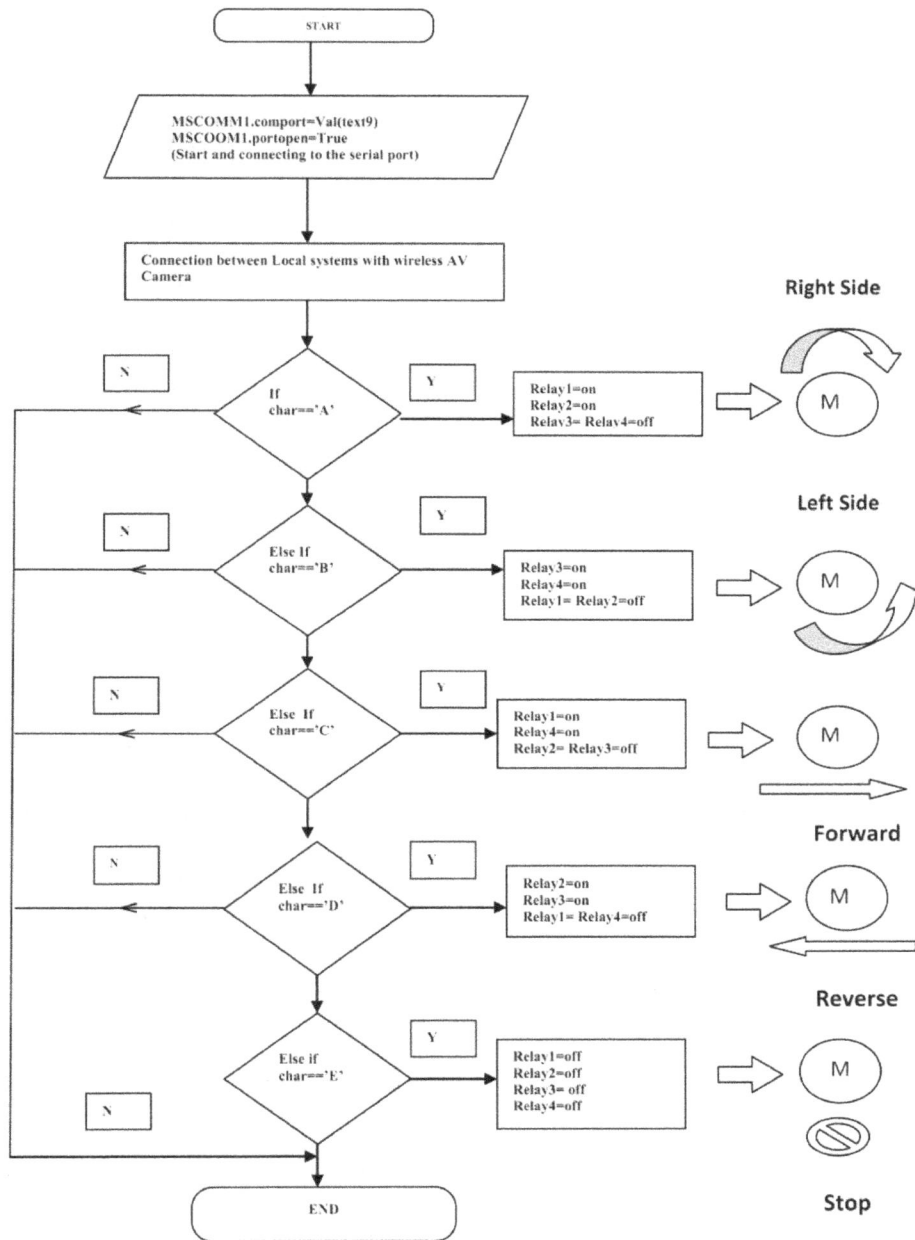

Figure 11: Flowchart for Robotic direction control

3.8 The Easy Cap USB 2.0 Grabber

The Easy CAP USB 2.0 Audio-Video adapter can capture high-quality audio-video files directed by USB 2.0 interface without sound card. However, the installation is very simple and the external power is unnecessary. It is the solution for laptop; we have enclosed the professional video editing software Tin Cam which provides best editing function. High-speed rendering and real-time performance ensures less waiting time and more time to create. Figure 13 and 14 shows the TV tuner and the Easy cap USB2.0 Grabber.

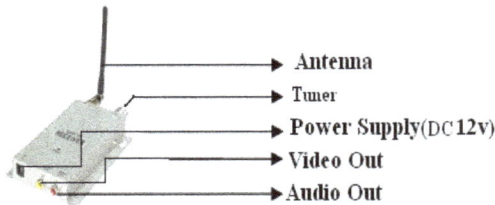

Figure 13: TV Tuner

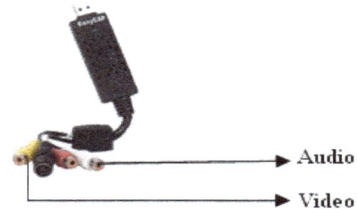

Figure 14: Easy Cap USB 2.0 Grabber

3.8.1 Key Features

- Popular USB 2.0 interfaces and does not need other power.
- Capture Video and Audio though USB 2.0 interfaces.
- Support Brightness, Contrast, Hue, and Saturation control.
- The dimension is suitable, that is easy to carry.
- Could capture audio without the sound card.
- High plug and play.
- Support For All Formats: record in DVD+/-R/RW, DVD+/-VR, and DVD-Video.
- Applying to internet conference and net meeting.

4. SOFTWARE DESCRIPTION

4.1. Embedded 'C'

Embedded C use most of the syntax and semantics of standard C, e.g., main() function, variable definition, data type declaration, conditional statements (if, switch. case), loops (while, for), functions, arrays and strings, structures and union, bit operations, macros, etc. In this project Keil cross compiler is used which is one such compiler that supports a huge number of host and target combinations. Use of C in embedded systems is driven by following advantages:

- It is small and reasonably simpler to learn, understand, program and debug.
- C Compilers are available for almost all embedded devices in use today, and there is a large pool of experienced C programmers.
- Unlike assembly, C has advantage of processor-independence and is not specific to any particular microprocessor/ microcontroller or any system. This makes it convenient for a user to develop programs that can run on most of the systems.
- As C combines functionality of assembly language and features of high level languages, C is treated as a 'middle-level computer language' or 'high level assembly language'.

4.2. Steps for Compiling, Linking and Downloading software in to the Microcontroller

Step 1: Create new project
Step 2: Select the device for the Target
Step 3: Add files to the source
Step 4: Build and Link the target program
Step 5: Compile and Debug the program
Step 6: Create Hex file from the source file
Step 7: Download the Hex file using willar programmer
Step 8: Hex file downloaded in to the AT89C51 successfully

4.3. VB front end

In this project visual basic 6 is used as a front end application at the local system and has been used to control the robot through Zigbee connection. This visual basic 6 front end application has mainly 3 forms each form contains some similar data and conditions. Form1 (Login page) is the login page for the security purpose, which provides the authentication to enter in to the next page for controlling the robot and without password any one can access and control the robot. And Form2 (control page) contains the information regarding robotic movement control and process the sensor information from the robot. According to the PIR and Gas sensor data, it displays the "Gas detected" and "Intruder detected" message at the front end application and immediately it calls the emergency form page (Form3). Form3 (Emergency alert page) performs the audio and visual alarm process for the security purpose. Figure 15 shows the VB front end application form wizard.

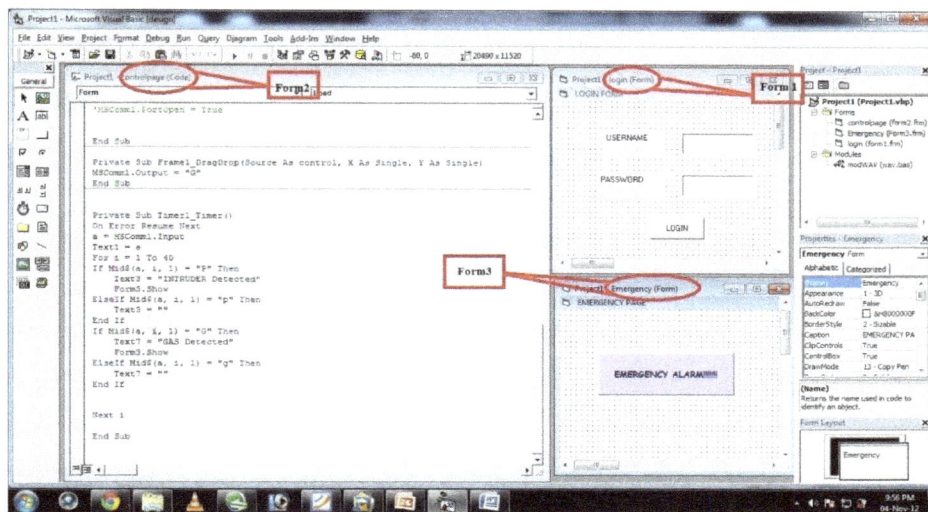

Figure 15: VB Front end application wizard

4.5 Tin Cam Software Specifications

Tin Cam can upload the pictures with FTP and simply save the pictures to a specified folder. The Windows Media Video format is used for streaming both audio and video signals. Tin Cam can create and upload a webpage that displays the webcam pictures or video stream. It allows writing text on the page, changing colors and adding a background pictures. Tin Cam can insert a caption on the webcam pictures, captions can contain time and date, and can be loaded from an external text file.

5. Results and Implications

The prototype of the robot is shown in the figure16. The robot has four wheels which are used for moving backward, forward, left and right turns. The Zigbee at the robotic side is used for transmitting the data from microcontroller to the local system [9]. The proposed robot is a battery powered and compact system. Fig 17 shows circuit diagram of the proposed system.

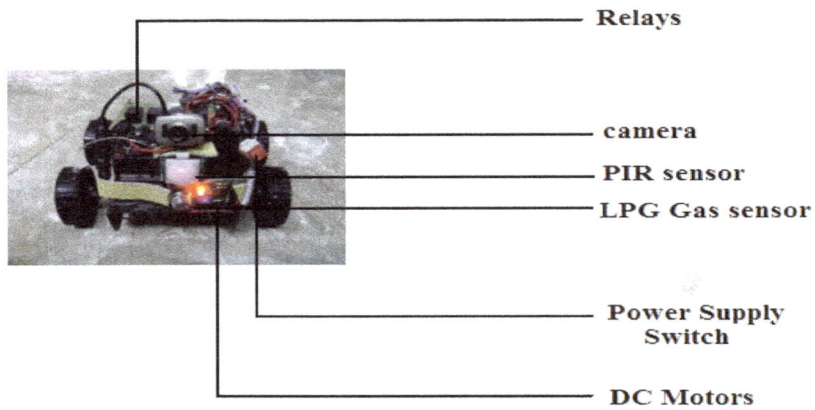

Figure 16: Prototype of the proposed system

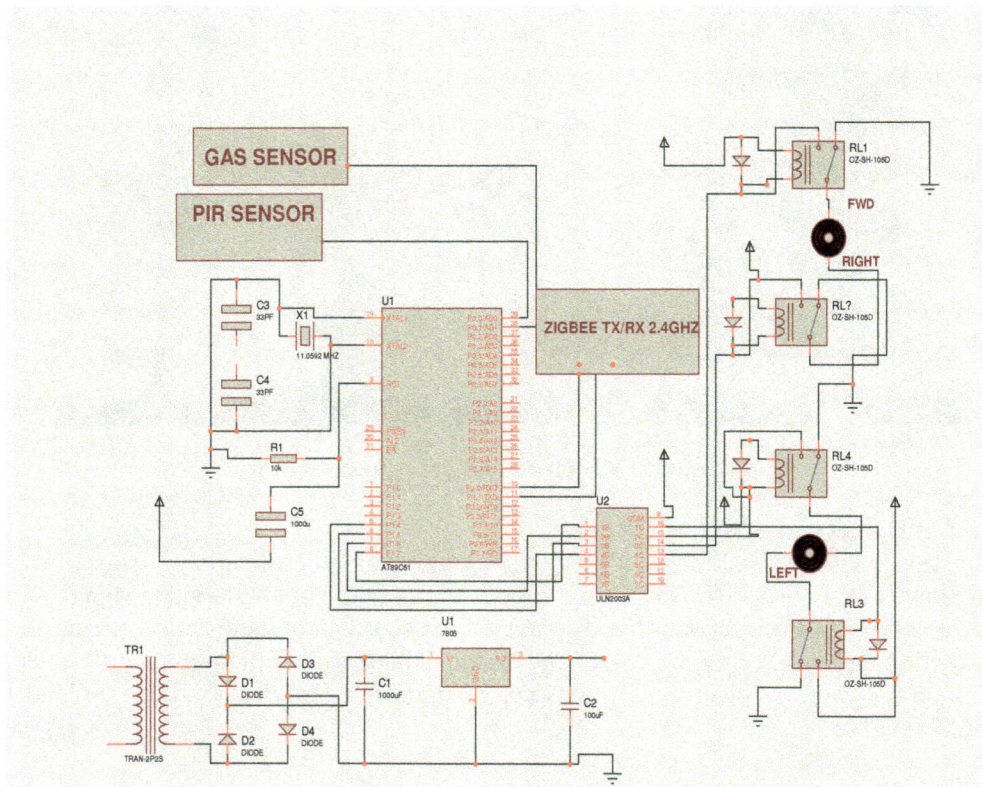

Figure 17: Circuit diagram of the proposed system

5.1 Audio-Video streaming and Robotic movement control at local system

The following steps shows the how the alert information are viewed at the local system on detecting the Gas and Intruder. And also shows the wizard for Audio-Video streaming.

Step 1: Login page

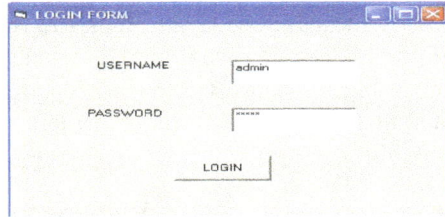

This wizard shows the login page at the local system, in which the authenticated user only can enter the username and password, and only if the password is correct it will show the following pages and allows controlling the robot.

Step 2: Gas Detection

Figure 18 shows the detection of LPG Gas, where the LPG Gas container is placed near to the Robot for the demonstration purpose [10]. The sensitivity for detecting the concentration scope for the bio-hazardous gases like LPG, iso-butane, propane and LNG are from 200 to 10000ppm. The Gas detection is indicated by red LED glow at the robot. Our Robot has detected the presence of LPG gas in its environment and we controlled and moved the robot near to the gas leakage source (Figure 18).

Figure 18: Gas Detection

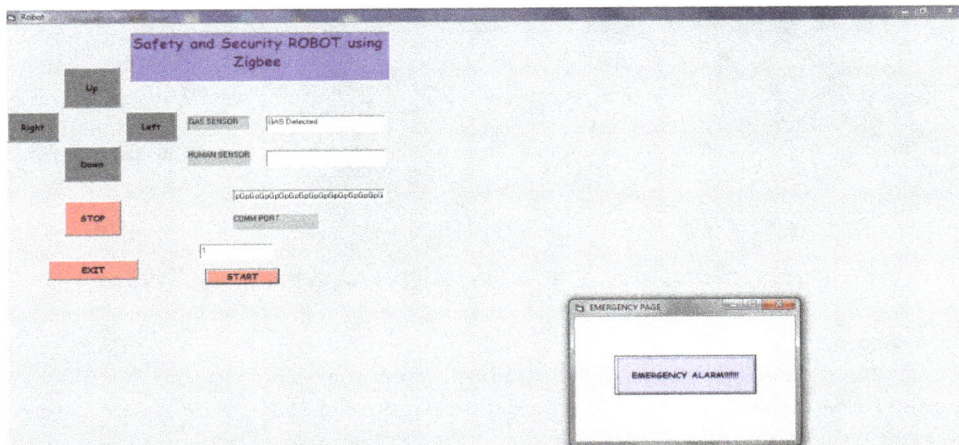

Implication 1:

The presence of gas is detected by the gas sensor, and then the same information is sent to the local system where it provides the audio and visual alert message which is shown in the above wizard. This alert information is highly helpful to avoid the disaster due to life threatening hazardous gas leakage. We have tested this in our chemistry lab, where we have chemicals and gases which are highly inflammable.

Step 3: Intruder Detection

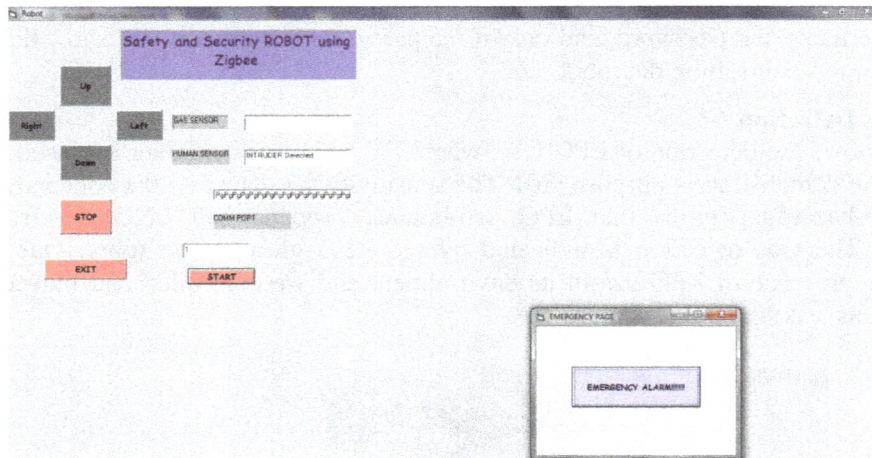

This window shows the intruder detection by the PIR sensor and provides alarm about the security status. TV Tuner is connected to the Local System through RF Wires with Easy Cap USB2.0 Grabber and transfers the Audio and Video signal from the wireless AV Camera as shown in figure 19. Zigbee is connected to the Local System through RS232 9 pin interface.

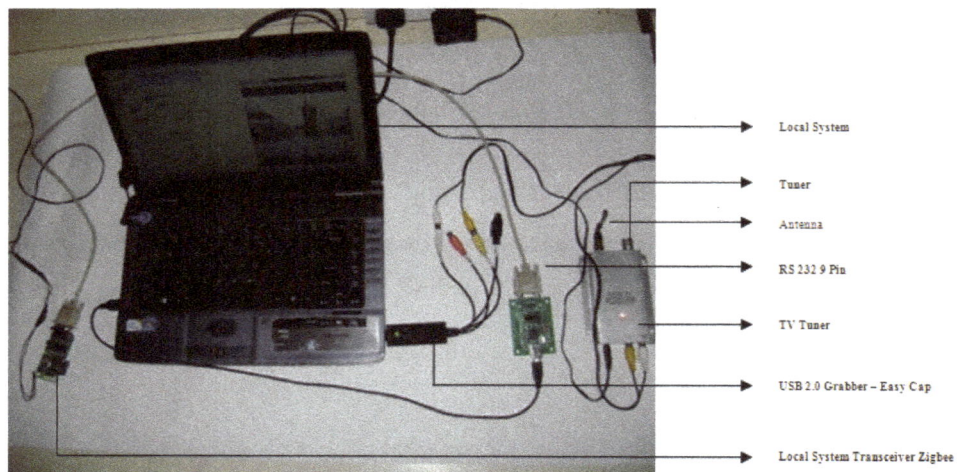

Figure 19: Interfacing TV tuner and Zigbee with the Local System

Implication 2:

We have communicated the sensor information from the robot to the local system and controlled the robot through zigbee communication (figure 7 and 19). The zigbee is operated on battery power with sleep mode which reduces the battery use. There is less probability of interfering with other users, since it incorporates CSMA-CA protocol and also supports automatic retransmission of data which ensures the robustness of the network. The distance between the robot and the local system supported by zigbee is 400ms only and hence the distance can be further more increased by using zigbee-pro which can support several kilometers depends on the cost. It facilitates ease-of-use and supports larger networks that comprised of thousands of devices. Our system supports interoperation with each other (zigbee and zibee-pro) and ensures long term use and stability.

Step 4: Audio-Video Streaming on Tin Cam

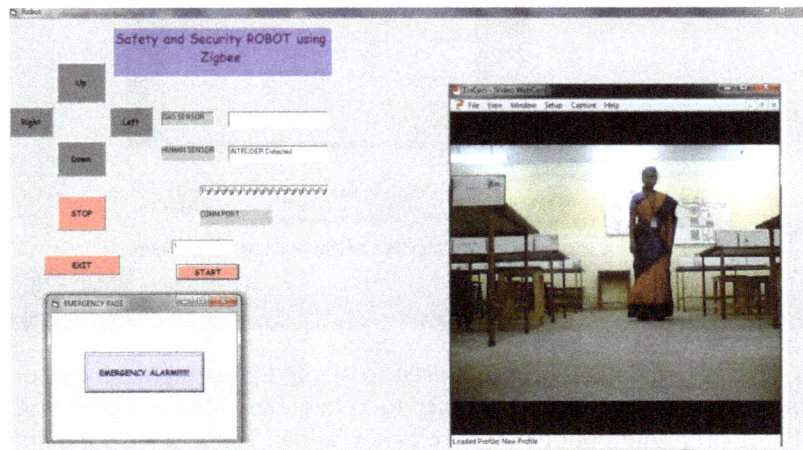

This page shows the video and audio buffering on Tin Cam software. We can easily adopt both windows in a same window. So it is very helpful to know the recent status of robot and detections of audio and video from wireless AV camera.

Implication 3:

In intruder detection, we have tested the security feature in our computer laboratory. The PIR sensor present in the robot senses the presence of intruder in a particular area and immediately provides a security alarm at the local station (Step-3 wizard). The low cost wireless AV camera at the robot captures its in-front audio and video information and performs AV streaming at the local system through Tin cam (step-4 wizard). This system is really helpful where security is a major threat. And also be useful to detect the people who were alive in the disaster situations to save the valuable human lives where we cannot go and find them.

5.2 Configuring Tin Cam with FTP server.

The following steps are the main procedure to upload the audio and video in FTP web server through Tin Cam. These steps are used to configure the Tin Cam with FTP server.

Step 1: Select USB 2.0 Grabber for connecting with the AV Wireless Camera driver interface card (EASY CAP). It can be viewed by clicking after setup menu and select "setupwizard" then enter the FTP connection details to update the video and audio to FTP Server, the process is shown in the following wizards.

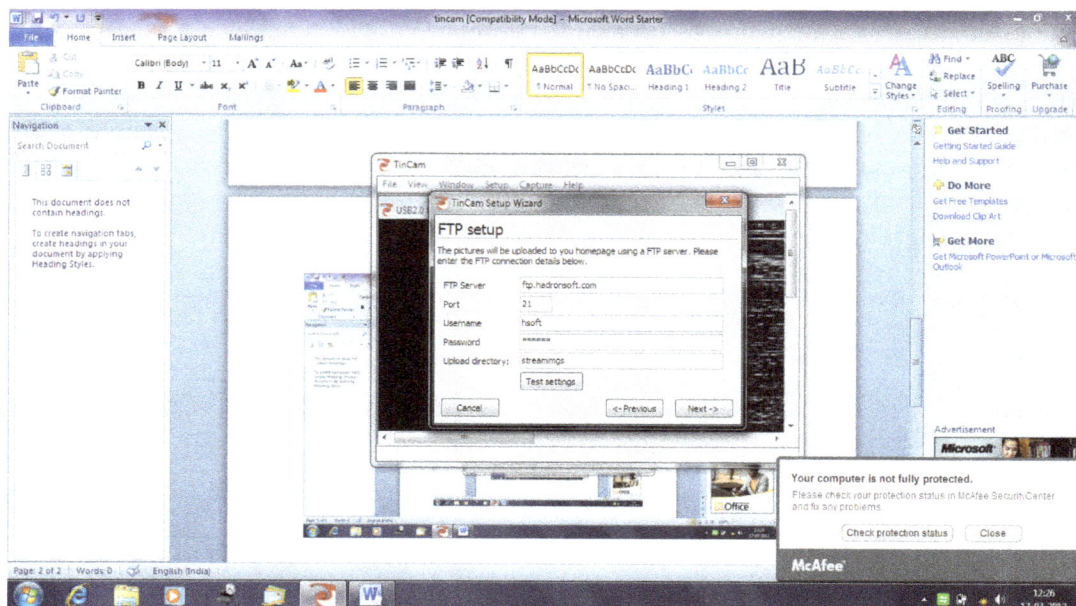

Step 2: The wizard shown below is used to create the web page by us or the software. Tin Cam can easily create its own web page to upload the video and audio on FTP server. Then Select the appropriate options that are given below. Here we select Tin Cam to create and upload the video and audio through webpage. The number of frames per second to buffer can be assigned by us according to the speed of our network.

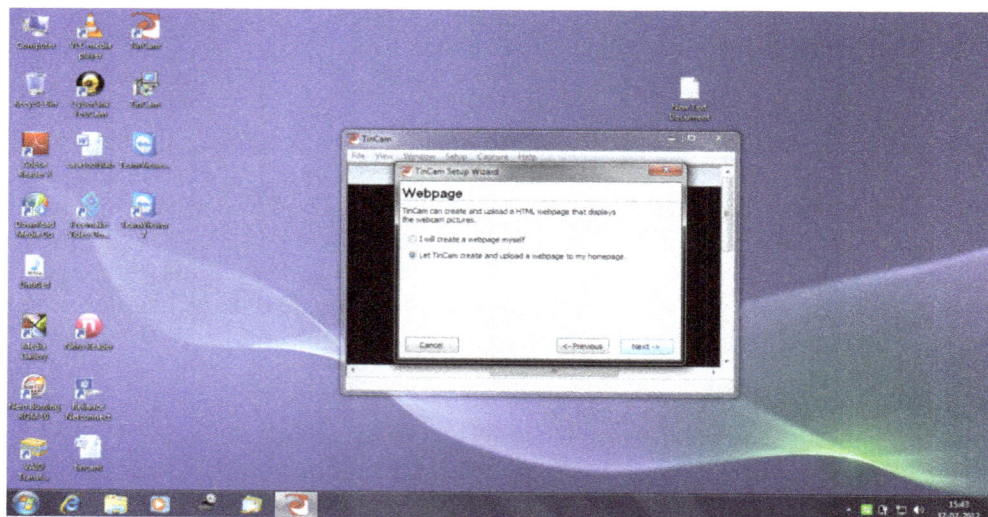

Step 3: Check the pointer that select USB 2.0 Grabber and insert the local System's IP address in the required box. To know the IP address of the Local System, go to command prompt and type "**IPCONFIG**". The IP address should be correctly typed without misspelled in IP address bar and deselect the "Auto detect" button. If you are having doubt in your network you can re-test your connection by clicking the "Test network connection" button. After that go to capture menu, then select Video Streaming Server to upload the video and audio streaming at the webpage also at the FTP web server.

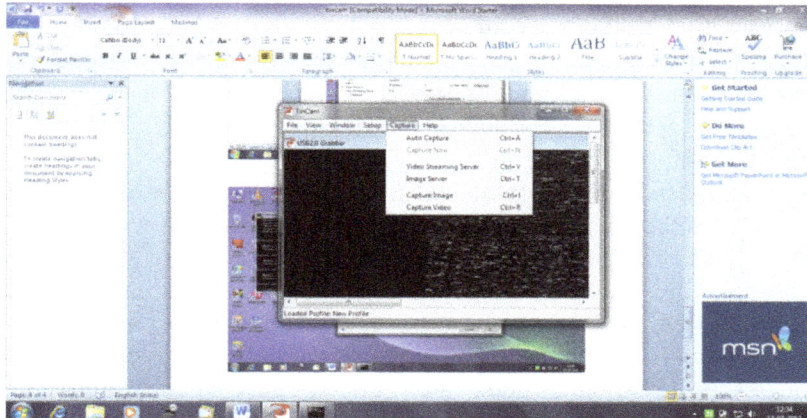

Step 4: After finishing the entire above setup wizard, then go to internet explorer or any relevant web browsing application then type this URL address "hadronsoft.com/streamimgs/webcam.html" to display the video and audio information at robot environment. Figure 20 shows the Audio and Video streaming at the web server and figure 21 shows AV streaming on 'N' number of Remote system.

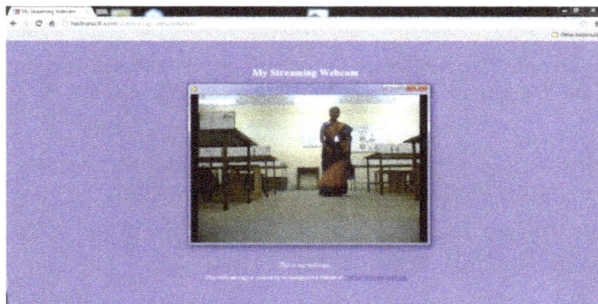

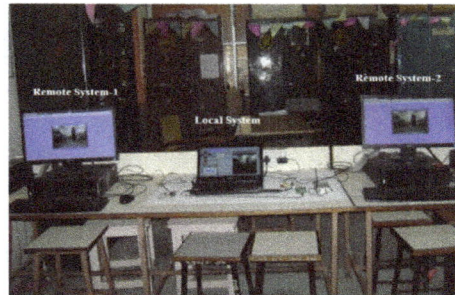

Figure 20: Audio-Video streaming at Web server

Fig 21: Audio-Video streaming on 'N' Number of Remote system

Implication 4:

We have tested and performed video streaming both in the local system and 'N' number of remote systems (figure 21). So any user can easily access the audio and video streaming at anywhere in the world. The tin cam supports the user in selecting the number of frames per second and if we increase the number of frames then we can get good resolution in AV streaming. In our system we select five frames per second (step 2). Tin cam also supports different protocols like TCP/IP, ISP, HTTP and FTP. We have uploaded the audio and video signals to the FTP web server (step 1, 2 and 3) through tin cam (figure 20). And hence this gives higher priority for security purpose.

6. CONCLUSION

The proposed robot can be used in war field, mines, power station, military operations, industries, research and educational institutions and so on. And also be used wherever people cannot go or where things doing too dangerous for humans to do safely. The Robotic movement is controlled remotely through the local system. The presence of bio hazardous gases like LPG, iso-butane, propane, LNG and alcohol were detected through MQ6 Gas Sensor which is placed at the robot. Similarly the intruder (Human or Animal) entered into the room/ range is detected through the PIR sensor. The above two sensed parameters were sent to the local system through the zigbee communication which is presented at both the ends, that is at the robot and at the local system. And at the same time an audio and visual alarm is raised [11]. A wireless AV camera resides at the robot; send's the robotic environment information to the local system. The video streaming is simultaneously done at both the local and 'N' number of remote system (web server). TV tuner is the source to receive the video signals from wireless AV camera and send that signals to the Local system using RF wires. The videos are streamed using the Tin Cam software, this software is used to create a web page to do live streaming through the web server. This system can be used where ever the safety and security are the major threat. In future this work may be enhanced in such a way that, whenever a picture is captured then a Tin Cam can immediately send an email about the picture. And also zigbee-pro may be used to increase the communication distance between the Robot and with the local system.

REFERENCES

1. V. Ramya, B. Palaniappan, and Subash Prasad, " Embedded Controller for Radar based Robotic Security Monitoring and Alerting System", International Journal of Computer Applications (IJCA), Volume 47-No. 23, June 2012.

2. V.Ramya, B.Palaniappan, K.Karthick and Subash Prasad "Embedded System for vehicle cabin toxic gas detection and alerting", Journal of Elsevier Procedia Engineering, 30(2012).

3. V. Ramya, B. Palaniappan , "Embedded Technology for Vehicle cabin safety Monitoring and Alerting System", International Journal of Computer Science Engineering and Applications, Volume 2-No.2, April 2012.

4. V. Ramya, B. Palaniappan , "Embedded system for Hazardous Gas detection and Alerting", International Journal of Distributed and Parallel Systems (IJDPS) Vol.3, No.3, May 2012

5. Heng-Tze Cheng, Zheng, Pei Zhang, "Real-Time Imitative Robotic Arm Control for Home RobotApplications", Carnegie Mellon University, IEEE, March 2011.

6. Hsian-I Lin, Yu-Cheng Liu, "Evaluation of Human-Robot Arm Movement Imitation", Nat.Taipei university of Technol., Taipei, Taiwan, IEEE, May 2011.

7. M.Gao, F.Zhang, andJ.Tian, "Environmental monitoring system with wireless mesh network based on embedded system,"inProc.5thIEEE Int. Symp. Embedded Computing, 2008, pp. 174-179.

8. Xia, F.; Sun, Y.X. Control and Scheduling Codesign: Flexible Resource Management in Real Time Control Systems; Springer: Heidelberg, Germany, 2008.

9. Ben Gaid, M.; Kocik, R.; Sorel, Y.; Hamouche, R. A methodology for improving software design lifecycle in embedded control systems. In Proc. of Design, Automation and Test in Europe (DATE), Munich, Germany, March 2008.

10. Ruijie Zhang Funjun He, Zhijiang Du and Lining Sun, "An Intelligent Home Environment Inspecting Robot," vol.42, pp. 140-169, 2007.

11. A.Cherubini, "Development of a multimode navigation system for an assistive robotics project", IEEE International Conference on Robotics and Automation, Rome, Italy, 10-14 April 2007.

12. L. Ma, C. Cao, N. Hovakimyan, C. Woolsey, V. Dobrokhodov, and I. Kaminer. Development of a vision-based guidance law for tracking a moving target. In AIAA Guidance, Navigation and Control Conference and Exhibit, August 2007.

13. K. Galatsis, W. Wlodarski, Y.x. Li and K. Kalantar-zadeh, "Vehicle cabin air quality monitor using gas sensors for improved safety," pp. 143-164, 2000.

6

The Development of a LAN for DVB-T Transmission and DVB-S Reception with Designed QAM Modulators and COFDM in the Island of Mauritius

Sheeba Armoogum[1], Vinaye Armoogum[2] and Jayprakash Gopaul[3]

[1]Department of Computer Science and Engineering, University of Mauritius, Reduit, Mauritius
s.armoogum@uom.ac.mu
[2]Department of Industrial Systems Engineering, University of Technology Mauritius, Port-Louis, Mauritius
varmoogum@umail.utm.ac.mu
[3]Mauritius Broadcasting Corporation, Reduit, Mauritius
suraj.gopaul@yahoo.fr

ABSTRACT

This paper is a thorough study of a digital television broadcasting system adapted to the small mountainous island of Mauritius. A digital TV LAN was designed with MPEG-2 signals. The compressed signals were transmitted using DVB-T and QAM modulators. QAM-16 and QAM-64 modulators were designed and tested with a simulator under critical conditions of AWGN and phase noises. Results obtained from simulation have shown that Digital video broadcast with a single frequency network (SFN) is possible in Mauritius with QAM-64 and QAM-16 modulators applying COFDM mode of transmission. However, this study has also shown that QAM-16 modulator had a better performance at low AWGN values (less than 12 dB) and can be adopted for Mauritius Island, provided that the number of transmitted channels is not high enough.

KEYWORDS

COFDM, Digital Television Broadcasting System, DVB-S, DVB-T, QAM

1. INTRODUCTION

Today, the management of a broadcast station is impossible without the use of computer-based technologies. At the onset, these technologies merged the world of radio, television and telecommunication. Now, they are creating new values, new possibilities and new services. This means that these new tools are contributing to the growth of an 'information society', in the broadest sense of the term [1]. The real revolution is being instigated by the packetizing of digital programme signals and the distribution of these packets in the form of data streams over internal and external networks. This will have a tremendous impact on radio and television and will entirely change their future form. In the studio, new computer-based tools are allowing us to store our programme content on servers, and to access or distribute this content via IT-based networks [2]. All cumbersome hardware equipment is now being replaced by software, that is, by something that can easily be changed and updated over time. This aspect is extremely important. These mammoth changes are, of course, leading to completely new workflows, particularly in the broadcast environment [1].

In 1999, COFDM (Coded Orthogonal Frequency Division Multiplexing) technology was implemented, and it subsequently provided the first step in support of the migration to fully digital transmission system architecture. The implementation and modulator architecture facilitate the interface to most types of digital transport streams, e.g. asynchronous serial interface (ASI), which is used to carry a compressed digital video bit stream [3]. The advantages of COFDM technology are apparent in its ability to offer error free transmission under severe multi-path conditions and its ability to occupy less overall bandwidth than its previous analog FM counterpart [3,4,5]. A typical COFDM waveform occupies 8 MHz of bandwidth at its 1dB bandwidth points [3]. Its digital implementation is based on the ETSI standard EN300-744 for DVB framing, channel coding and modulation architecture [5]. The uniqueness of the standard allows the user to customize his data throughput as a function of three main parameters: forward error correction, guard interval, or delay spread and modulation type [6].

Following international trends in the way broadcasting technologies are evolving, the local broadcast station has no choice other than to walk on the paste of these changing technologies [7]. Maintenance of Beta Cam analogue equipment is becoming more and more costly as parts are becoming rare in the international market. However, swapping to new digital technologies is not an easy task. The implementation of new digital broadcasting techniques is being done phase by phase over some period of time. At the broadcasting station, new digital equipment has been installed in parallel to the old analogue equipment. Personnel are being trained to work on the new equipment. At this stage of digitalization, lots of problem is becoming apparent. There is no compatibility between the old format of transmission and the new one. Digitalization and networking in broadcasting house concerns four mains department (figure 1), namely the News production (editing), production department (editing), Archives and preview unit, Television transmission and satellite reception room.

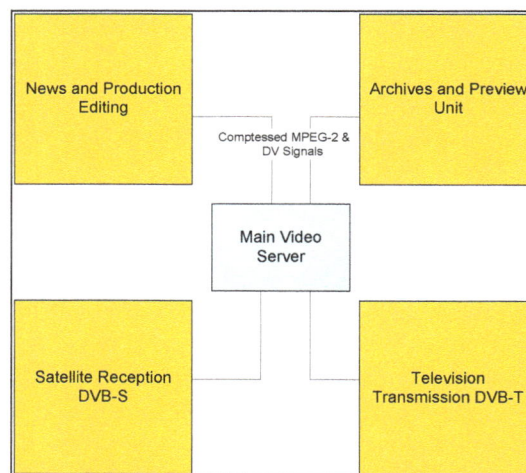

Figure 1. Block diagram of the 4 mains department networked over the main Video Server

This research is mainly concerned with audio and video signals for Television transmission, Archiving and Editing. Compressed MPEG-2 video signals are received from satellite source. The signals are demodulated using QPSK and converted to analogue before being stored in Beta Cam or DV Cam magnetic tapes. At the transmission level, analogue signals processed at the broadcast house are sent to Multicarrier Ltd (a third party organisation) for broadcast over multiples frequencies network. With the increase in number of channels, it is becoming more and more difficult for regulatory bodies to allocate bandwidth for transmission. On the other hand, digital transmission with sophisticated and robust modulation techniques is occupying limited bandwidth for the same number of transmission channels. In 2005, MBC has launched its DVB-T channels. But still, with the lack of digital networking for audio and video transmission, reference signals sent to Multicarrier Ltd, in analogue form, is processed to digital at the transmission station and aired. This results in loss in signals quality and freezing of picture frames due to high intersymbol interference (ISI) at the modulator [4].

Path loss analysis, Quality of service for the design of broadcast terrestrial network and the setting up of appropriate equipment at base station are now vital. People want to receive good quality of audio and video, mainly when the world is moving towards the digital era. Many researchers over the years since the launching of digital television have conducted research in these areas. Fernandez et al. have studied indoor digital radio reception in the Medium Wave band [8]. Heuck has derived a model of a hybrid network and performance analysis of hybrid networks is discussed [9]. In 2005, Perez-Vega and Zamanillo have done intensive research works on derivation of models and evaluation of the quality for digital broadcast network [10]. Arinda et al. [11,12] have done similar works in Spain on digital TV broadcasting. We have carried some preliminary study on field strength with few measuring data in the north [13,14, 15] and have studied the variation of the path loss using some models [16]. This paper is the continuation of research works done recently by us since 2005.

The aims of this paper are firstly to design a local area network (LAN) for multimedia capture, storage and transmission, secondly to design modulators for DVB-S at the satellite reception and DVB-T modulator for Terrestrial transmission and lastly to study the COFDM for digital DVB-T transmission for television. The objectives therefore will include optimizing the LAN network and video server, studying and comparing different modulation techniques such as QPSK, QAM64, and QAM16 and finally simulating signals for the modulators and choose for the most efficient modulator for DVB-S and DVB-T signals.

2. DESIGN OF A LAN FOR DIGITAL BROADCASTING

Conventional TV transmission standards are based on technology that is more than 40 years old, and the world has been split into three discrete television systems – NTSC, PAL and SECAM – that are incompatible with each other and difficult for international program exchange [2,17,18,19]. As a result there has been widespread interest in the electronics industry to develop more advanced TV systems, including digital high definition television. Digital TV refers to digital representation and processing of the signal (Digital LAN) as well as digital transmission (DVB-T) and digital reception (DVB-S) [4,7].

Digital TV broadcast involves converting images and sound into digital code. This digitization of images and sound (data) starts with compression, in order to minimize the capacity requited of transmission channels (bandwidth).Digital TV transmission standard uses processing and compression to achieve simultaneous transmission of several different television programs [5]. The quality of signal received is equivalent to the studio output.

Digital television represents a dramatic change for the production and broadcast industries, as well as the users. New technologies have recently brought tremendous flexibility in the use of different pictures format using digital compression systems. Moreover, because of digital nature of the picture information and the emergence of powerful high-speed processors and graphic cards, the computer industry is directing its main business to the broadcasting world. These converging technologies are modifying completely the existing TV environment.

2.1. The Need for Local Area Media Network

As discussed earlier, media network at the Mauritius Broadcasting station (MBC) concerns four main departments, namely the editing suites (News and production), preview unit, television broadcast studios and the satellite reception room. Since the creation of the station in the year 60's, heavy and bulky equipment have been utilized for image capture, processing and transmission. It's only by the end of the 90's that MBC started to invest on software based equipment and low-cost workstations for editing and transmission. Following this trend and with major changes in Broadcasting technologies (Beta cam to DV), the station has acquired numbers of workstation with powerful software (Avid non-linear editing) for editing and broadcast [18]. The editing suites are not interconnected together in a network (stand alone), and video from one editing station is copied (dubbing) on magnetic tape and transported to other departments like preview unit or television. There is loss in video quality (generation) due to numerous dubbing. The price of standard magnetic tape for video broadcast is soaring up the sky. Archiving of uncompressed videos in magnetic tape is becoming more and more difficult. Prices for Beta Cam hardware equipment are so high that maintenance of existing machines is becoming a burden. On the other hand, the prices of powerful media workstation and software are becoming more accessible. The cost of high storage hard disk (Terabytes) is much less than that of magnetic tapes of same storage capacity. For example the cost of storing 5000 hours of media on magnetic tapes is much more than that of high quality compressed media (MPEG, DV, etc) on hard disk.

The advantages of having most of the workstations grouped in a network are
- Easy sharing and transferring of data, and protection of the data
- Application sharing
- Easy interaction of other users in the network
- Sharing of peripherals devices

As shown in figures 2, 3, 4 and 5, the proposed model for the networking architecture interconnecting the four main departments inside the Broadcasting House is presented. Details of the architectures are described in the next sub-sections.

2.1.1. Server network connections

As shown in figure 2, the main server is a database server of around 16 terabytes of uncompressed media storage capacity. This will serve mainly the purpose of storing video files in MPEG formats as well as AVI and OMF and DV. The stored media will be used by editing suites and television for transmission. Data from satellite feed, being compressed (real time video streaming) in MPEG-2 layers are multiplexed and sent to the database on real time basis. The server is programmed in such a way that live feeds Medias from satellite can be stored for four consecutive days before being override applying a first in first out basis. The Video server (figure 2) is the heart if the network as it will communicate with all end users.MPEG-2 video files on the server will be utilized by the preview units where they are sorted and selected prior to editing and transmission.

The Video database server will be interconnected to the preview workstations, Satellite room computers, Transmission room workstations, Editing server and a Linux proxy server for internet purposes. Here also a star physical topology is adopted with a 64 ports switch connected to all the workstations and to the 8 port switch of the LANShare server throughout a gateway.

As depicted in figure 3, the gateway used can be a CISCO switch, Pix Firewall or Symantec hardware firewall with 4 ports entry [20]. This network is connected with a Symantec hardware firewall. The hardware firewall is chosen instead of the CISCO switch because of security aspect due to internet utilization. Symantec hardware firewall of same properties costs less than CISCO Pix firewall. However one advantage of Symantec firewall is that it protects individual workstation on the network from virus attack, Denial of service attack (DOS attack) privileges and can be authenticated. In practice, most computers have designed network security. Disk-based video storage is generally recognized as being more flexible than tape, particularly when considering the transfer of Data Essence and Metadata. Disks permit flexible associations of Data Essence and Metadata with the video and audio. A video server has the characteristic that it views Content from both a network / file system perspective and a video / audio stream perspective and it provides connectivity to both domains. Constraints are derived from internal server performance and from the transport / interface technology that is used when conveying the data to / from the video storage system. Due to the currently-used interfaces, there are limitations on the way that manufacturers have implemented video servers but many of these limitations will be removed as interfaces develop. The manufacturers of disk-based video storage will typically support a wide variety of formats, interfaces and Metadata capabilities. It is recommended that the manufacturer's video storage specifications include Storage capacity, Sustained data transfer, Video Essence types supported, Audio Essence types supported, Data Essence types supported, Corrected BER, Interfaces supported and Metadata and Data Essence recording limitations as the essential parameters [5].

Within the studio there are likely to be several different classes of server, corresponding to the principal work activities within the studio – in particular, play-out servers, network servers and production servers (figure 4). These will differ in storage capacity, streaming capability and video-processing capability. Here, we deal only with the issues of transportation and storage of Data Essence and Metadata. It is important that the formats of Data Essence and Metadata employed be consistent throughout the production, distribution and emission chain. It is not desirable to track identical Data Essence and Metadata with a number of different identification schemes. In the case of Metadata, a mechanism has been established through the Metadata Registration Authority. All suppliers and users are recommended to use this facility. It is further recommended that all Essence formats (Data, Video streams and Audio streams) be registered through the same authority. Users should be able to share data electronically between databases. These databases may contain either Data Essence (e.g., subtitle text) or Metadata information. Users should not be required to manually re-enter this data at any stage beyond first entry. Indeed in many instances, Metadata and some types of Data Essence ought to be automatically created. There is also a requirement for database connectivity between broadcast and business systems for the exchange of data such as scheduling information and operational results. This connectivity should be handled in a standardized way as far as possible.

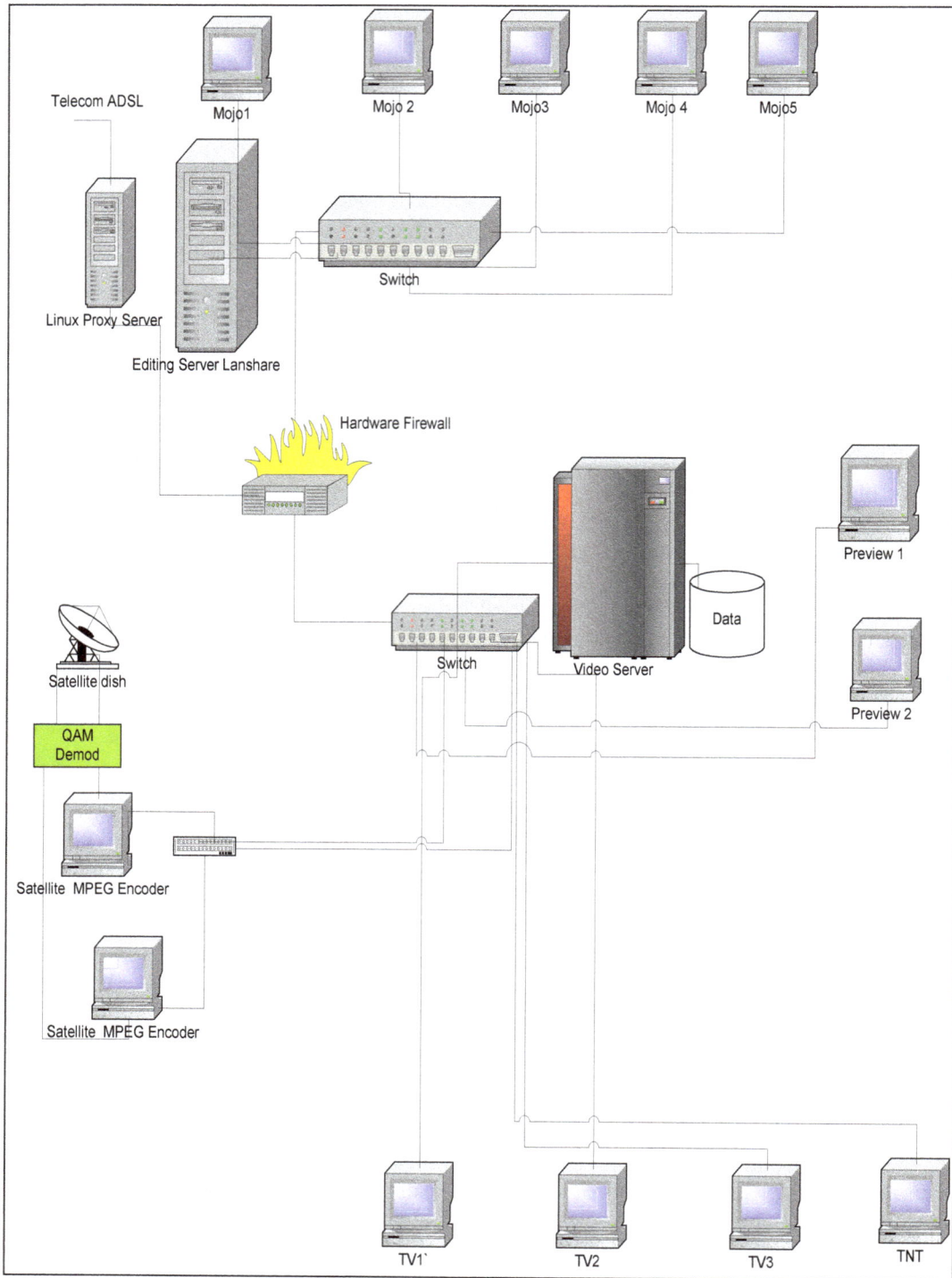

Figure 2. Proposed architecture for Digital LAN

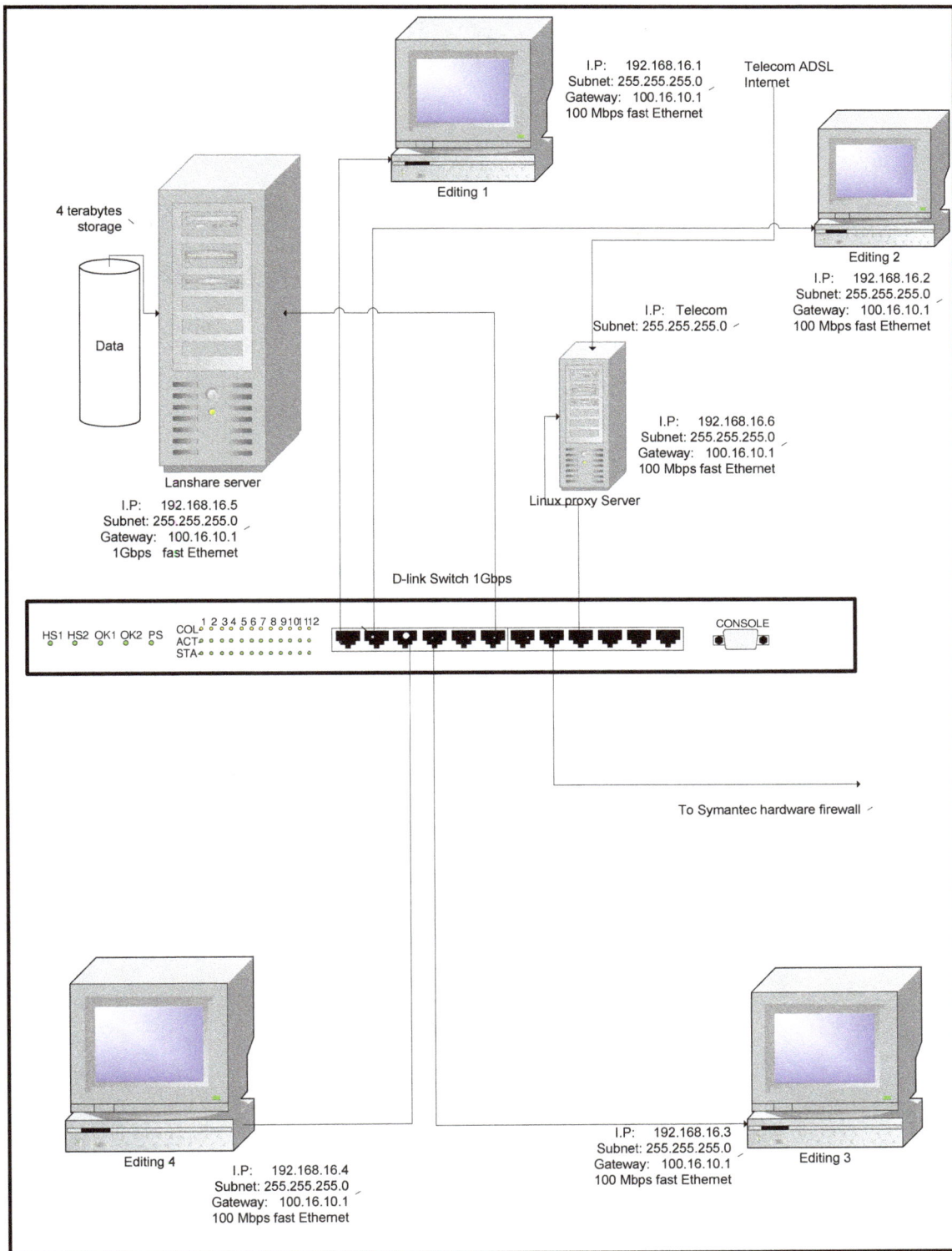

Figure 3. Detailed network design with IP addresses

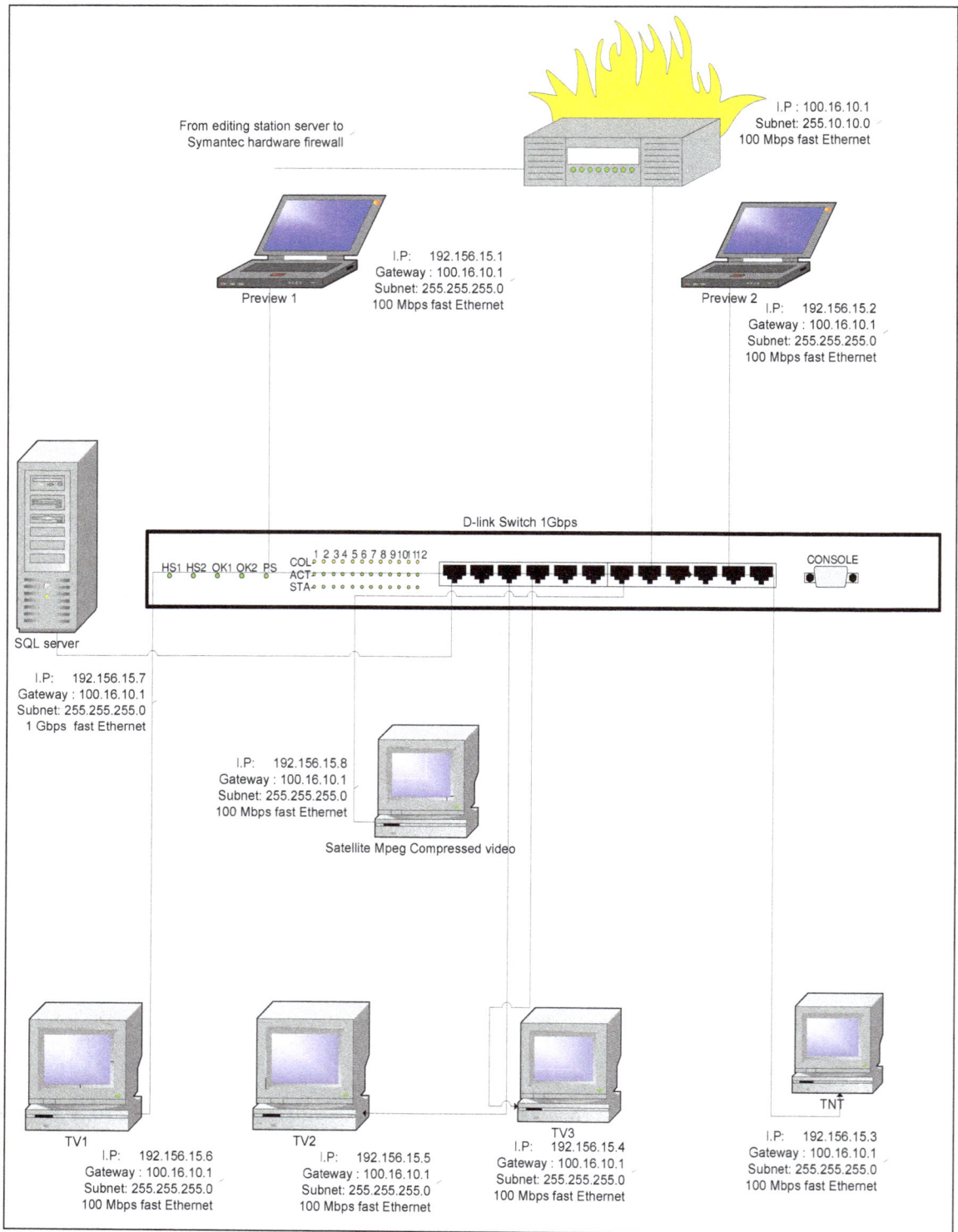

From editing station server to
Symantec hardware firewall

I.P : 100.16.10.1
Subnet: 255.10.10.0
100 Mbps fast Ethernet

I.P: 192.156.15.1
Gateway : 100.16.10.1
Subnet: 255.255.255.0
100 Mbps fast Ethernet

Preview 1

Preview 2

I.P: 192.156.15.2
Gateway : 100.16.10.1
Subnet: 255.255.255.0
100 Mbps fast Ethernet

D-link Switch 1Gbps

HS1 HS2 OK1 OK2 PS   COL
ACT
STA

1 2 3 4 5 6 7 8 9 10 11 12

CONSOLE

SQL server

I.P: 192.156.15.7
Gateway : 100.16.10.1
Subnet: 255.255.255.0
1 Gbps fast Ethernet

I.P: 192.156.15.8
Gateway : 100.16.10.1
Subnet: 255.255.255.0
100 Mbps fast Ethernet

Satellite Mpeg Compressed video

TV1

I.P: 192.156.15.6
Gateway : 100.16.10.1
Subnet: 255.255.255.0
100 Mbps fast Ethernet

TV2

I.P: 192.156.15.5
Gateway : 100.16.10.1
Subnet: 255.255.255.0
100 Mbps fast Ethernet

TV3

I.P: 192.156.15.4
Gateway : 100.16.10.1
Subnet: 255.255.255.0
100 Mbps fast Ethernet

TNT

I.P: 192.156.15.3
Gateway : 100.16.10.1
Subnet: 255.255.255.0
100 Mbps fast Ethernet

Figure 4. Detailed network design and configurations

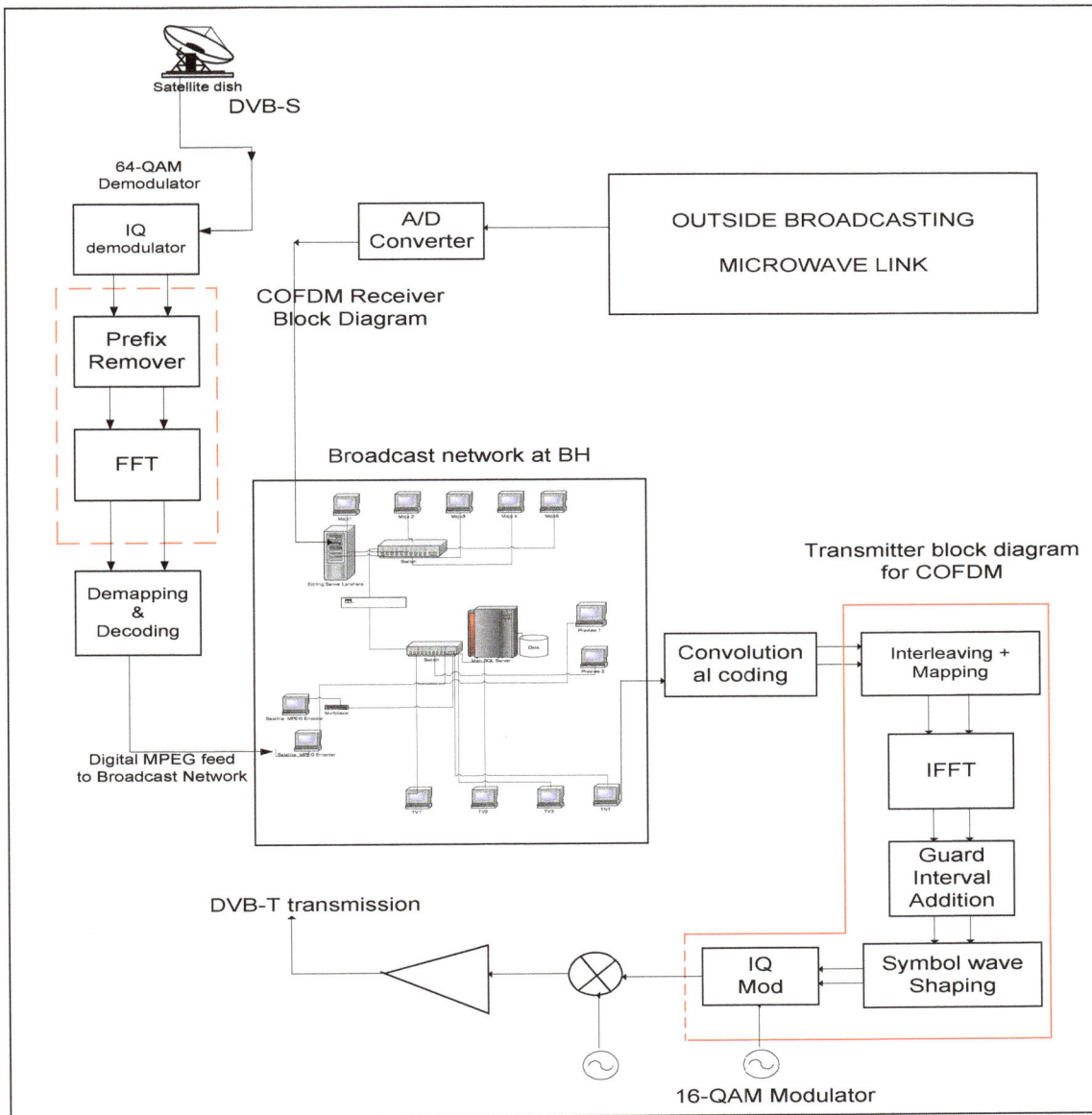

Figure 5. Entire network design and configurations

2.1.2 Firewall

In the network circuit designed for MBC broadcast station, a Symantec hardware firewall (figure 2) is installed and it acts as a gateway for incoming internet signals and the editing server. The firewall can be programmed to allow and block access to any workstation in the network. The firewall can be administered by an administrator using any workstation on the circuit. The hardware firewall works with a 100 Mbps network card thus allowing fluid traffic flow throughout the network. Connections to the ports are made using RJ-45 socket and 1Gbps cable having standard code IEEE 802.3. Rather than directly forwarding the incoming packets, the proxy firewall uses a special proxy application to generate fresh packets – based on the incoming request – onwards from the other network.

2.2. The Implementation of COFDM System for DVB-T Transmission and DVB-S Reception

As shown in figure 5, Coded Orthogonal Frequency Division Multiplexing (COFDM) has been specified for the digital broadcasting system for both audio - Digital Audio Broadcasting (DAB) - and (terrestrial) television - Digital Video Broadcasting (DVB-T) as well as satellite broadcasting (DVB-S). COFDM uses a very different method of transmission to older digital modulation schemes and has been specifically designed to combat the effects of multipath interference for mobile receivers. COFDM can cope with high levels of multipath propagation, with a wide spread of delays (Guard Spaces) between the received signals. This leads to the concept of single-frequency networks in which many transmitters send the same signal on the same frequency, generating "artificial multipath". COFDM also copes well with co-channel narrowband interference, as may be caused by the carriers of existing analogue services.

2.2.1. Concatenated coding system for DVB-T transmission

As depicted in figures 5 and 6, Coding of bits (Signal) is achieved into four major phases
- The input data bits are first converted to parallel by using serial to parallel block.
- Vector Reed-Solomon forward error coding is then performed
- The encoded data is then passed through a frequency interleaver where 223x8 bits symbol converted by the RS-encoder to 255-symbol are arranged in arrays where a pattern of 8-bit per block is formed.
- The signal is further encoded (inner coding) by applying fractional rate coding to further performance enhancement in the presence of frequency selective fading and interference.

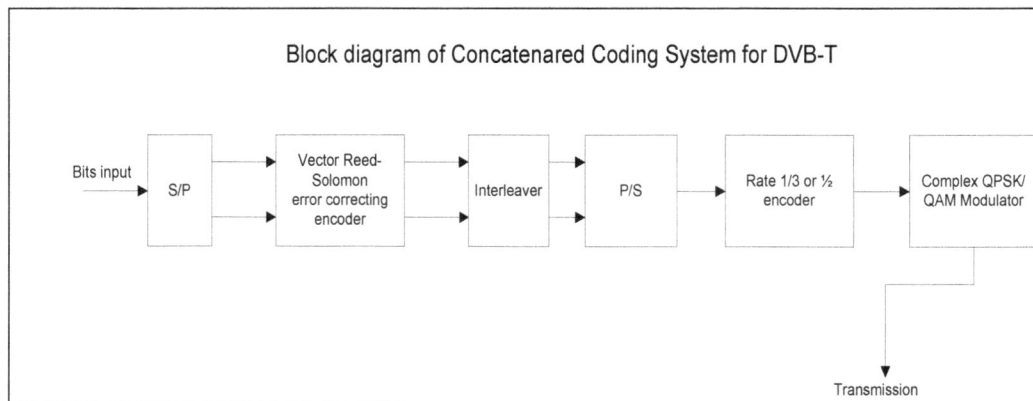

Figure 6. Concatenated coding system for DVB-T

2.3 Modulation Techniques

Modulation is the process of encoding information from a message source in a manner suitable for transmission. It generally involves translating a source signal (baseband) to a bandpass signal at frequencies very high compared to the baseband frequency. The bandpass signal is called the modulated signal and the baseband message signal is called the modulating signal. Modulation can be done by varying amplitude, phase or frequency of high frequency carrier in accordance with the amplitude of the message signal. The benefit of using 16-QAM or 64-QAM is that each symbol on each subcarrier can carry more bits of information. Of course, it is better to use a higher level constellation so that the overall capacity can be higher, but the drawback is that the points are closer together which makes the transmission less robust to errors. Fading alters both the amplitude and phase of a carrier or subcarrier, and in the mobile channel the frequency of the subcarriers are altered by a Doppler shift. However, analyzing the three modulation techniques explained, it can be said that a 16-QAM modulator having 2 subchannels and with protection distance larger than 64-QAM would perform better in the presence of frequency-selective fading environment.

3.4. OFDM system of transmission

The OFDM block (figure 7) is transformed into the time domain by means of IFFT (Inverse fast fourier transform or IDFT). Guard interval is added to the signal to prevent inter symbol interference. In the receiver, the guard interval is removed and the signal is transformed into frequency domain by means of Fast Fourier transform (FFT). After demodulation bits are mapped applying soft decision values and then iterative decoding is performed by the vertebi decoder to obtain the output data.

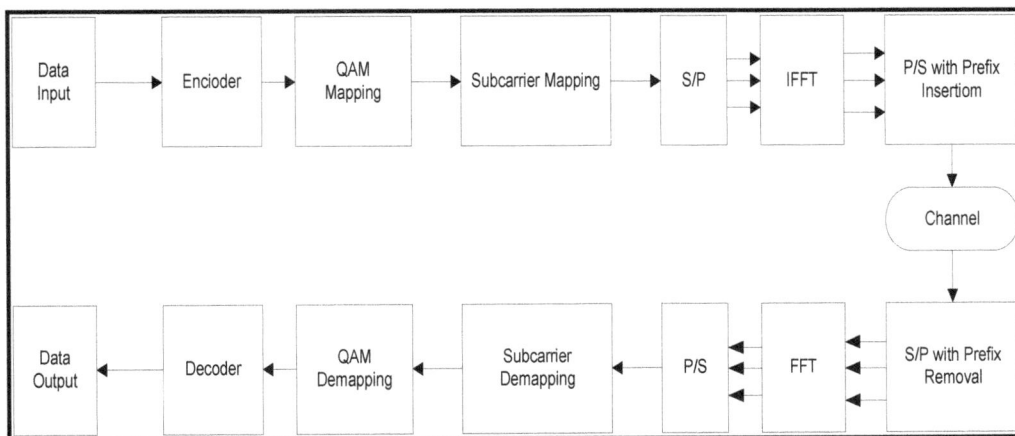

Figure 7. OFDM system of transmission

3.5. Transmission and reception system over designed Architecture

Figure 5 is a complete representation of the proposed digital network which encompasses the DVB-T system of transmission with a 16-QAM modulator, and a DVB-S system of satellite reception with a 64-QAM demodulator. The transmitted signal is first passed through the RS encoder for forward error correction, and then it is fractionally coded through the Vertibi encoder (addition of punctured convolution code). After interleaving the signal it is then converted into a parallel stream before being transformed into the time domain by the inverse Fast Fourier transform (IFFT). The signal is then mapped into the 16-QAM modulator with a capacity of 4 bits per symbol. After addition of guard interval; the modulated signal leaves the station as COFDM.

At the Satellite reception, the COFDM signal is digitally tracked, and then the signal is demodulated using a 64-QAM .The demodulated signal is then passed through the guard interval remover. The parallel signal is then converted back into the frequency domain by applying the Fast Fourier transform (FFT). The signal is then converted into serial form before being decoded by the Vertibi and the RS decoder.

An additional source of input to the broadcast network is the outside broadcast signal. Signal from outdoor is transmitted via microwave link; it is captured and compressed into digital form before being stored into the video server.

3.6. Proposed design for DVBT transmission over Mauritius

Figure 8 is therefore the proposed architecture for DVBT transmission over Mauritius. COFDM signals are modulated using QAM modulators. MPEG-2 compressed video is sent for transmission. The channel is divided into frequency spacing of 200 MHz and are orthogonally coded. The generated n random bits are encoded using a rate-3/4 punctured convolutional encoder. The resulting vector contains 4/3n bits, which are mapped to values of -1 and 1 for transmission. The puncturing process removes every third value and results in a vector of length 8/9n. The punctured code, punctcode, passes through an additive white Gaussian noise channel. After passing through a 12x12 array frequency interleaver, the signals is sent to the QAM mapper where it get modulated for transmission with appropriate guard interval.

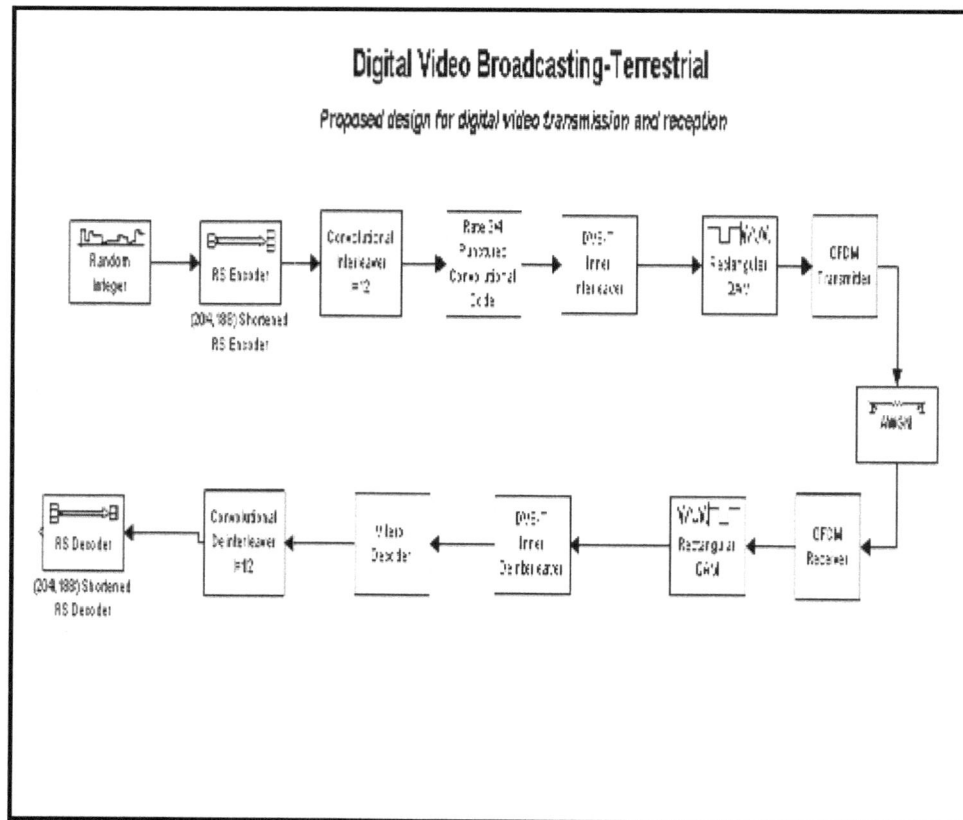

Fig.8. Proposed design for DVB-T transmission in Mauritius

4. RESULT AND DISCUSSION

To perform analysis QAM modulators and demodulators are simulated from MATLAB. An AWGN channel adds white Gaussian noise to the signal that passes through it. The relative power of noise in an AWGN channel is typically described by quantities such as:

(i)   Signal-to-noise ratio (SNR) per sample. This is the actual input parameter to the AWGN function
(ii) Ratio of bit energy to noise power spectral density (Eb/N0).
(iii) Ratio of symbol energy to noise power spectral density (Es/N0)

4.1. Comparison of BER and SER for different modulators

An analysis of the bit error rate and the symbol error rate is performed while increasing the bit energy to noise power density (Eb/N0). As depicted in figures 9, 10 and 11, the graph of error probability is plotted against Eb/No (dB) for QPSK, QAM-16 and QAM-64. It is observed that as the number of bits per symbol is increased, the symbol error rate and bit error rate increase accordingly from an ordinary QPSK to QAM-64. Since the SER is more affected than the BER, it can be said that all the three modulators can be utilized for DVBT transmission at Eb/No in the range 10-15 dB. But as for QPSK, which can rotate limited amount of bits at a time, it will cause long delay in signals. QAM16 and QAM64 are therefore better for digital video broadcast.

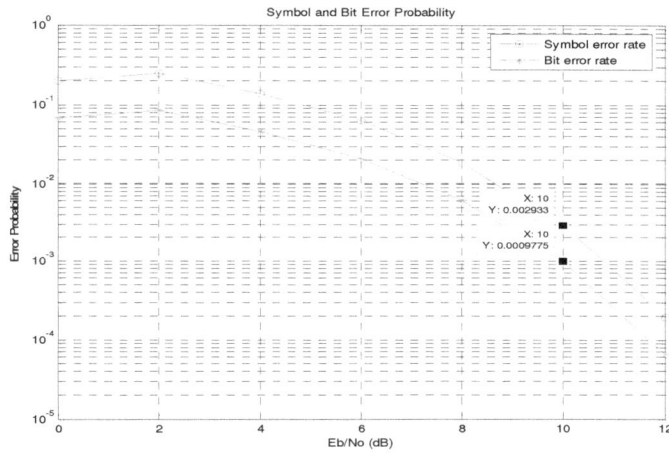

Figure 9. Graph showing symbol error rate and bit error rate for QPSK modulator

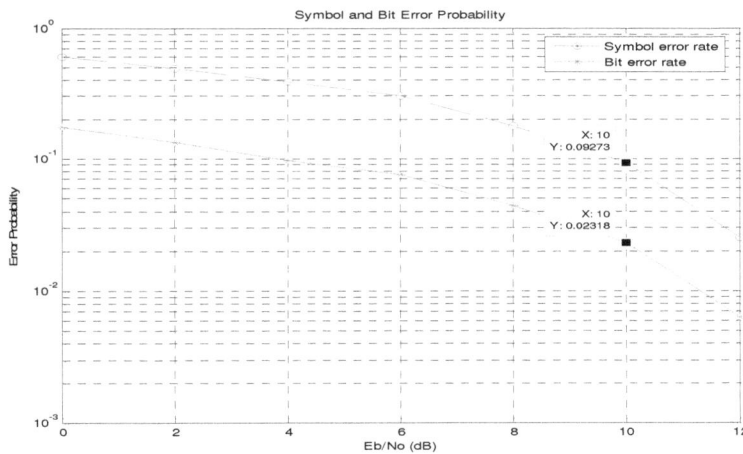

Figure 10. Graph showing symbol error rate and bit error rate for QAM-16 modulator

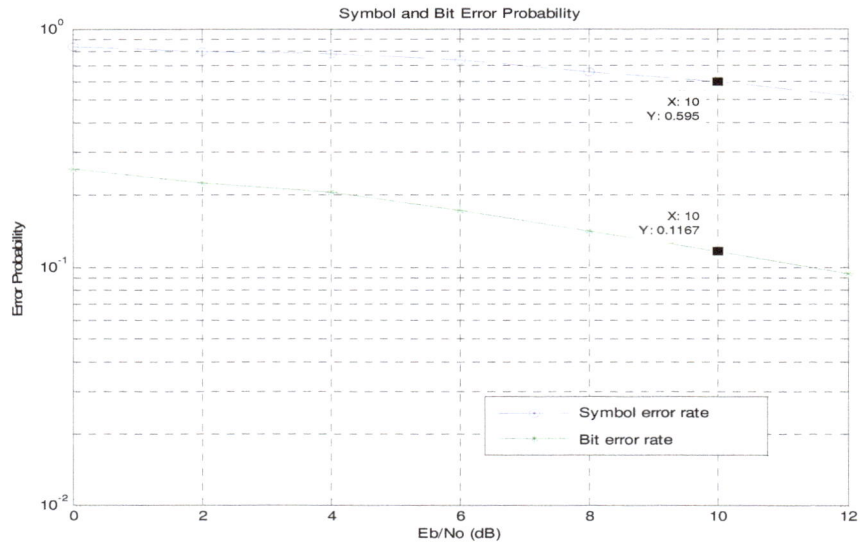

Figure 11. Graph showing symbol error rate and bit error rate for QAM-64 modulator

4.2. BER Tools analysis

When ratio of bit energy to noise power spectral density (Eb/N0) is set to 0-18 dB, the following outputs are observed (figure 12). Taking phase noise and AWGN into consideration and assuming an average of 15 dB of bit energy to noise density as an extreme condition, we can observe that there is a subsequent increase in the bits error rate (BER) from the QPSK model to the QAM16 and QAM64 model.

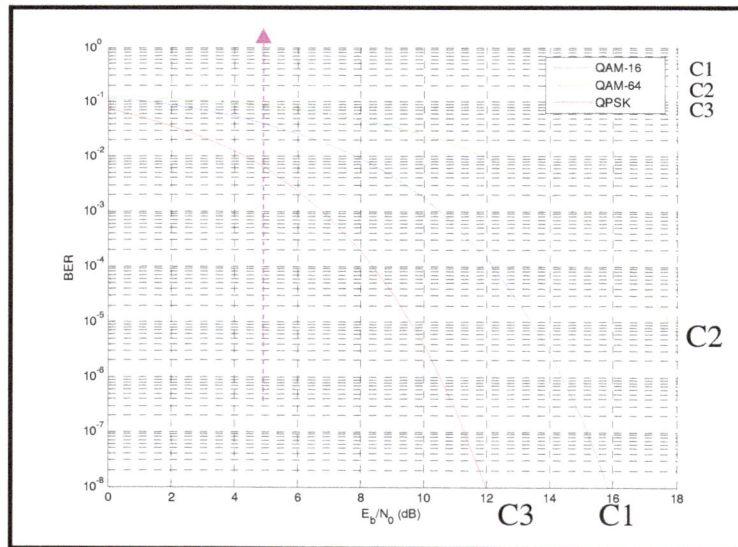

Figure 12. Graph representing the variation of BER with Eb/No

However, at 15 dB the bit error rate can still be tolerated as the respective modulators shows that the bit error rate will still be of small magnitude of 10^{-3} .They are within the norms set up by European and American standards for broadcasting.

4.3. Simulations with noise addition

The main idea of this simulation is to compare two types of signal. The first one is the original signal sent for transmission and the second one is the transmitted signal. The transmitted signal is modulated, added to white Gaussian noise and phase noise, and then demodulated. The difference between the pure signal and the transmitted signal is given by the error rate calculator. Finally the total number of symbol errors is displayed.

The blocks and lines in the model describe mathematical relationships among signals and states:
   (i)   The Random Integer Generator block, labeled "Random Integer," generates a signal consisting of a sequence of random integers between zero and 255
   (ii) The Rectangular QAM Modulator Baseband block, to the right of the Random Integer Generator block, modulates the signal using baseband QAM or QPSK modulators.
   (iii) The AWGN Channel block models a noisy channel by adding white Gaussian noise to the modulated signal.
   (iv) The Phase Noise block introduces noise in the angle of its complex input signal.
   (v)   The Rectangular QAM Demodulator Baseband block, to the right of the Phase Noise block, demodulates the signal.
   (vi) In addition, the following blocks in the model help to interpret the simulation:
   (vii)   The Discrete-Time Scatter Plot Scope block, labeled "AWGN plus Phase Noise," displays a scatter plot of the signal with added noise.
   (viii)   The Error Rate Calculation block counts symbols that differ between the received signal and the transmitted signal.
   (ix) The Display block, at the far right of the model window, displays the symbol error rate (SER), the total number of errors, and the total number of symbols processed during the simulation

The first simulation is performed with a 16-QAM modulator (figure 14) and demodulator units. The result obtained is tabulated (table 1) and a Q/I plot of the transmitted signal is displayed as shown in figure 13. The inter symbol interference is less in QAM-16.

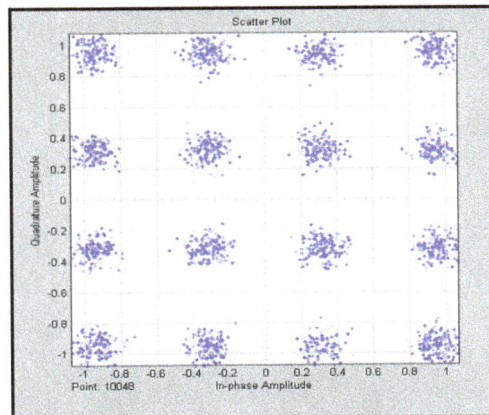

Figure 13. Scatter plot showing mapping of 10000 bits on QAM-16

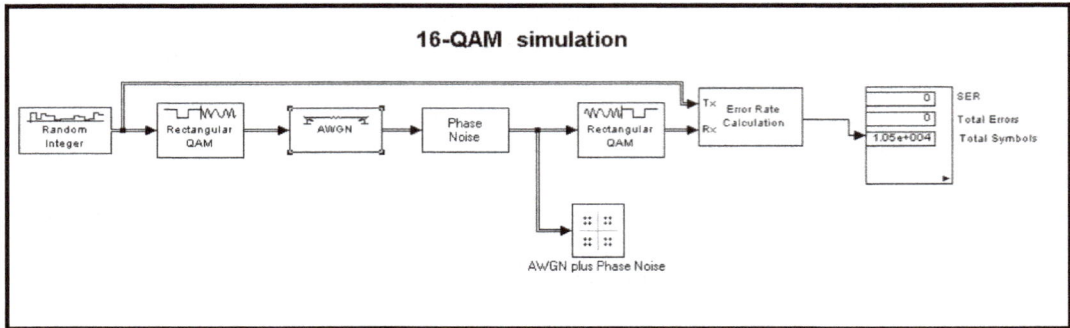

Figure 14. 16-QAM simulator

The same procedures are repeated for a 64-QAM modulators and demodulators (figure 16). The conditions of noises are kept constant. The Q/I plots of the 64-QAM and QPSK are displayed similarly (figure 15). A shown in figure 15, the inter symbol interference is more pronounced in QAM-64 compared to the result obtained as depicted in figure 13.

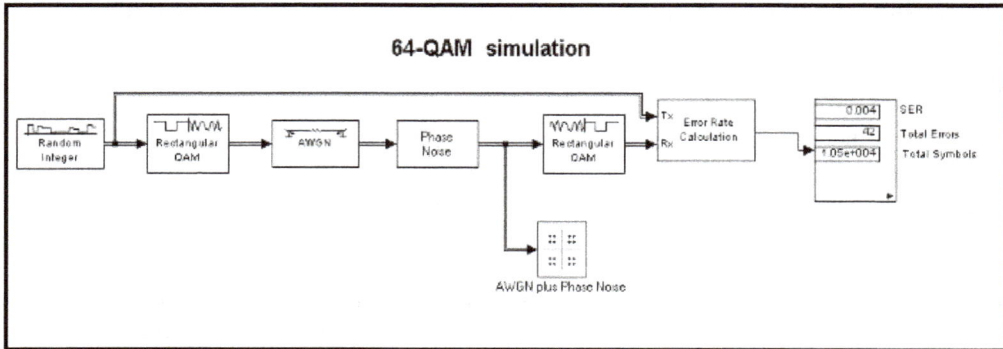

Figure 15. 64-QAM simulator

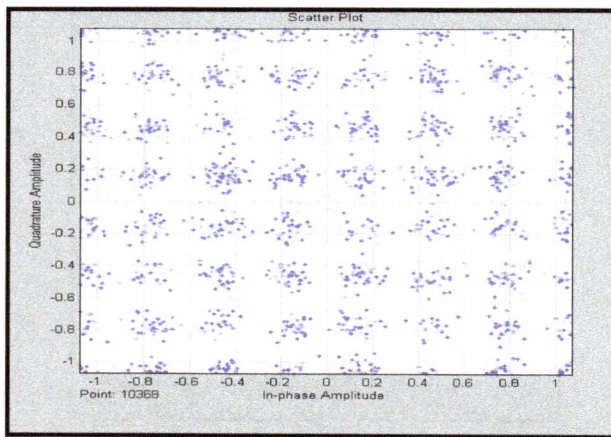

Figure 16. Scatter plot showing mapping of 10000 bits on QAM-64

Simulations are performed with the three types of modulators and the results are displayed as shown in the table below. For consistency of results, the phase noise value is kept constant at -76 dB/Hz and the total number of symbol processed is kept at 10500.The designed pi/4 QPSK simulator is also tested with 10500 symbols. The AWGN values vary from 12 dB to 15 dB and results for total number of symbol error out of 10500 symbols is obtained. Symbol error rate is also tabulated for the three modulators at different AWGN values (Table 1). The values are then compared with the standard values set by the European and American acceptable norms for broadcast.

Table 1. Simulation results for QAM/QPSK modulators

Modulator/ Demodulator	No of bits Per symbol	AWGN Eb/No(dB)	Phase Noise dB/Hz	Total Symbols	Total Errors	SER
QAM-16	4	12	-76	10500	4	0.000381
QAM-64	6	12	-76	10500	598	0.05695
QPSK	2	12	-76	10500	1	0.00095
Increasing AWGN from 12 to 15						
QAM-16	4	15	-76	10500	0	0
QAM-64	6	15	-76	10500	42	0.00400
QPSK	2	15	-76	10500	0	0

Results obtained from the phase noise simulator have shown that decreasing the ratio of bit energy to noise power spectral density (Eb/N0) or decreasing the Signal to noise ratio below 15 dB increases the symbol error rate of the QAM and QPSK modulators. However, keeping the phase noise value constant at -76 dB/Hz and changing the AWGN value from 12 dB to 15 dB gives rise to a decrease in the symbol error rate for the QAM-64 modulator. Keeping the Eb/N0 at 15 dB as threshold value and increasing further by 1 dB further decreases the symbol error rate. QAM-16 performance at an ideal condition of 13 dB is optimum. The error is 1 symbol over 10500 of processed symbol. After analyzing the results, it can be said that a QAM-16 and a QPSK modulator perform better at low SNR or Eb/N0. However, the time taken to process the 10500 symbol was fastest for the QAM-64 modulator followed by the QAM-16 Modulator. QPSK with 4 symbols per rotation took much more time to process the very high number of bits.

Digital video broadcast (DVB-T and DVB-S) involve highly compressed media and high number of bits. QPSK with four symbols per mapping would definitely not process the high volume of bits from the digital channel. There will be buffering of bits at the modulator and consequently the 'gel d'image'. QAM-16 with 16 symbols per rotation and accepting 4 bits per symbol gives the ideal solution for digital video broadcast; but here also it all depends on the number of channels to be sent to the modulator for processing.QAM-64 is the easiest solution for digital video broadcast provided that the terrain condition of Mauritius is good enough. QAM-64 has been adopted by MCML (Multicarrier Mauritius Ltd) the national broadcast for the transmission of terrestrial digital video broadcasting [13]. The performance of the modulator is good in some region and at stable weather condition. Mauritius being a tropical mountainous island with fast climatic changes needs a thorough study of the whole areas. Noise variations in different regions of the island should be analysed so that a precise indication can be obtained about the efficiency of QAM-64 modulation.

Finally, as it was shown in the first simulation that symbol error rate is much higher than bit error rate, we can deduce that at 15 dB of Eb/N0 satisfactory reception of digital signal can be achieved.QAM-64 is a good solution for DVB-S and can be used to demodulate the large number of channels from the satellite reception. For a small island like Mauritius surrounded by water and mountains, QAM-16 would be the right solution for DVB-T provided that the number of transmitted channels is reasonable.

5. CONCLUSION

Digital transmission and single frequency network are the future of video and audio broadcasting. Most of the well known international stations in the World have migrated towards digital broadcasting. Digital broadcasting has not only cut down the cost of broadcast equipment, but it has also created lots of space over the bandwidth of transmission. This is the right time for implementing digital transmission over Mauritius. Digitalization will increase the number of broadcast channels and as well increase the quality of sound and picture that is broadcast. Digital transmission and reception requires a well designed broadcast LAN.

In this paper we have given a thorough explanation about the requirements of a video LAN. The digital network was designed according to the norms set by the European broadcasting Union. A video server was connected over a star network with the help of CISCO switch and Gateway firewall. We have given a detailed study of COFDM and modulation techniques. After the detailed design of QAM and QPSK symbol constellations, we have noted that QAM modulators and demodulators are the right solution for digital video broadcast. We found that inter symbol interference was much less in QAM-16 than the in QAM-64 and also increasing the number of symbols meant increasing the number of bits to be transmitted. Results show that the utilization of QAM-16 modulator for DVB-T transmission throughout the island would be the most viable solution. QAM-16 is robust and resists fading effects even at an AWGN value set at Eb/No of 12 decibel. Inter Symbol interference is minimal even when more that 10000 symbols are mapped. ISI for QAM-16 is within the norms set by the European Broadcasting Union. It has been demonstrated from the Phase noise simulator that out of 10 500 symbols processed at AWGN 15dB, there was no symbol error. QAM-64 demodulator would be the right solution for DVB-S since satellite reception involves large number of digital channels and require modulators with higher processing speed .However, the symbol error rate of QAM-64 is higher at AWGN 15dB and the QAM-modulator is more likely to cause ISI and ICI. Finally, to conclude this research based on our findings it would be recommend that we adopt COFDM as a mode of transmission and that QAM-16 modulators be used to transmit compressed video and in the form of MPEG-2 over Mauritius Island.

FUTURE WORK

This research was carried out so as to develop a transmission system for digital video broadcast. Digital video broadcast is a wide field of study. Compression of signals, encoding signals and the way signals are transmitted varies a lot and can be the major field of study of any researcher. With the migration from MPEG-2 to MPEG-7 in a near future, lots of amendment will have to be made to the digital broadcast network. As an immediate future work, there is the need to study how a database system can be developed for the video server. Actually, in Mauritius, outside broadcasting is being carried out using microwave links. The future trend will be the development of an IP based transmission system over the designed digital network using fast ADSL line and router. At the transmission level further research will have to be done on the QAM modulators. We should study about possibilities of developing QAM modulators with layers of symbols. For example, QAM-16 modulator can be worked out with two layers of 16 symbols, thus allowing more channels to be processed and at the same time decreasing ICI and ISI caused by a single layer 64-QAM modulator

ACKNOWLEDGEMENTS

The authors would like to thank all the engineers and staff of the Mauritius Broadcasting Corporation (MBC), as well as the national broadcasting television carrier, the Multi Carrier Mauritius Ltd (MCML) for their tremendous help in terms of equipment and resources.

REFERENCES

[1]   P. Symes, (2001) "Video Compresion Demystified", Mc Graw-Hill, International Edition, ISBN 0-07-18964-5

[2]   T.S.Rappaport, (2005) "Wireless Communications Principles and practice", Prentice-Hall of India,Second Edition, ISBN 81-203-2381-5

[3]   J.H. Stott: The effects of phase noise in COFDM, EBU Technical Review, No. 276, Summer 1998.

[4]   C. Anderson & Mark Minasi, (1999) "Mastering Local Area Network", BPB Publications of India, First Indian Edition, ISBN 81-7656-060-X

[5]   ETS 300 744 (1997) "Digital broadcasting systems for television, sound and data services; framing structure, channel coding and modulation for digital terrestrial television", available at http://www.etsi.fr

[6]   Advanced digital techniques for UHF satellite sound broadcasting. Collected Papers on concepts for sound broadcasting into the 21st century, European Broadcasting Union, 1988

[7]   J.H. Stott, (1995)" The effects of frequency errors in OFDM, BBC Research and Development Report No.RD 1995/15", available at http://www.bbc.co.uk/rd/pubs/reports/1995_15.html [] Wendell Odom, (2001) "Cisco CCNA Exam certification guide", BPB Publications of India,First Edition, ISBN 81-7635-651-4

[8]   I. Fernandez, P. Angueira, D. De la Vega, I. Pena, D. Guerra, U. Gil, (2011). "Carrier and Noise Measurements in the Medium Wave Band for Urban Indoor Reception of Digital Radio". IEEE Transactions On Broadcasting , Vol 57, Issue 4

[9]   C. Heuck, (2010) "An Analytical Approach for Performance Evaluation of Hybrid (Broadcast/Mobile) Networks", IEEE Transactions On Broadcasting, Vol. 56, No. 1, pp. 9-10

[10]   C. Perez-Vega, J.M. Zamanillo, "Path Loss Model for Broadcasting Applications and Outdoor Communication Systems in the VHF and UHF Bands", IEEE Transactions On Broadcasting, Vol. 48, No. 2, pp. 91-96, 2002.

[11]   A. Arrinda, M. Ma Velez, P. Angueira, D. de la Vega, J. L. Ordiales, Digital Terrestrial Television (COFDM 8k System) Field Trials And Coverage Measurements In Spain, IEEE Transactions On Broadcasting, Vol. 45, No. 2, pp. 171-176, (1999).

[12]   A. Arrinda, M. Ma Velez, P. Angueira, D. de la Vega, J. L. Ordiales, Local-Area Field Strength Variation Measurements Of The Digital Terrestrial Television Signal (COFDM 8k) In Suburban Environments, IEEE Transactions On Broadcasting, Vol. 45, No. 4, pp. 386-391, (1999).

[13]   V. Armoogum, T. Fogarty, K.M.S. Soyjaudah, N. Mohamudally, (2006) "Signal Strength Variation Measurements of Digital Television Broadcasting for Summer Season in the North of Mauritius at UHF Bands", Conference Proceeding of the 3rd International Conference on Computers and Device for Communication, pp. 89-92.

[14]   V. Armoogum, K.M.S. Soyjaudah, N. Mohamudally, T. Fogarty "Height Gain Study for Digital Television Broadcasting at UHF Bands in two Regions of Mauritius", Proceedings of the 2007 Computer Science and IT Education Conference. Nov 2007 pp. 017-025, ISBN 9789990387476.

[15]   V. Armoogum, K.M.S. Soyjaudah, N. Mohamudally, T. Fogarty "Path Loss Analysis between the north and the south of Mauritius with some Existing Models for Digital Television Broadcasting for Summer Season at UHF Bands", Proceedings of the 8th IEEE AFRICON 2007 –ISBN 0-7803-8606-X.

[16]   V. Armoogum, K.M.S. Soyjaudah, N. Mohamudally, T. Fogarty "Comparative Study of Path Loss with some Existing Models for Digital Television Broadcasting for Summer Season in the North of Mauritius at UHF Band", IEEE The Third Advanced International Conference on Telecommunications (AICT-07), ISBN 0-7695-2443-0.

[17]   EBU technical Review (2003), available at http://www.ebu.ch/en/technical/trev/trev_295-poland.pdf

[18]   Jochen Schiller, (2003) " Mobile Communications ",, Pearson Education(Singapore) Pte. Ltd, ISBN 81-7808-170-9

[19]   Gaudrel R., Betend C., DIGITAL TV BROADCAST Field Trials on the Experimental Network of Rennes, Internal VALIDATE document from CCETT, FT.CNET/DMR/DDH, November 1997.

[20]   Wendell Odom, (2001) "Cisco CCNA Exam certification guide", BPB Publications of India,First Edition, ISBN 81-7635-651-4

Authors

Sheeba Amoogum received her BSc in Maths, Physics and Electronics in 1997 and Master in Computer Applications in 2000. She is currently doing her PhD. She is a lecturer at the University of Mauritius. Her fields of study are the software engineering, VoIP and networking.

Vinaye Amoogum received his BSc (Eng) in Computer Engineering and MSc (Eng) in System Engineering in 1995 and 1997 respectively. He completed successfully his PhD in Telecommunications in 2009. He is currently a senior lecturer at the University of Technology, Mauritius. His fields of study are the Telecommunications and related areas, computer science and engineering.

Jayprakash Gopaul is currently System Analyst/Administrator at the Mauritius Broadcasting Corporation. He has just completed an MSc degree in computational Science and Engineering in 2007.

7

Multiple Criteria Clustering of Mobile Agents in WSN

Yashpal Singh[1], Kamal Deep[2] and S Niranjan[3]

[1]Research Scholar, Mewar University, Rajasthan, India
yashpalsingh009@gmail.com
[2]Assistant Professor, Department of Electronics and Communication ,JIET
Jind,Haryana,india
erkdjangra@gmail.com
[3]Professor, Department of Computer Science & Engineering ,PDM College of
Engineering Bahadurgarh ,Jhajjar,Haryana,India
niranjan.hig41@gmail.co

ABSTRACT

In Wireless sensor networks data aggregation with hundreds and thousands of sensor nodes is very complex task. Recently, mobile agents have been proposed for efficient data dissemination in sensor networks. In the traditional client/server based computing architecture, data is collected from multiple sources and forwarded to destination for further processing. It requires high bandwidth, whereas in the mobile agent is a task specific executable code traverses to the relevant source for gathering data. It reduces communication overhead, reduce cost, low bandwidth. Agents have capability to perform task for multiple applications. It will send only useful information to destination node. The problem is to group similar mobile agents into a number of clusters such that each cluster has similarity in responding to a group of nodes. By clustering intelligent mobile agents, it is possible to reduce the cost of time for each individual agent, decrease the demand imposed on network for a set of required tasks, decrease total number of visits. This paper, we present the problem of Multiple Criteria Clustering of Mobile Agents (MCCMA) where the decision is to cluster mobile agents such that a group of similar intelligent mobile agents will visit a group of similar sensor nodes.

KEYWORDS

Clustering, wireless sensor network, power consumption, mobile agents.

1. INTRODUCTION

Wireless Sensor Network (WSN) has received much attention in the research field in last few years. Sensors are expected to be inexpensive and can be deployed in a large scale over the network of operation. A fundamental concept of WSN is that all sensor nodes are distributed in specific area and them remains fixed in their respective position. Sink sensor node gather information about environment, compute it and then send it to clients. Energy-efficient data delivery and security is crucial as sensor nodes operate with limited battery power. Due to their flexibility and cost effectiveness, wireless sensor networks (WSNs) have been used for numerous applications including environmental monitoring, facility monitoring, and military surveillance for tasks such as target detection[17]. Although there has been an extensive research work done toward energy efficiency of WSN. In the traditional client/server based computing model, information exchange takes place amongst sensor nodes acting as source and sink node. In case a linked bandwidth of WSN is low and cannot meet network traffic and require consumption of more energy. A lot of research has been done on power utilization in WSN. Various researchers have identified different routing algorithms, scheduling algorithms, load balancing techniques, clustering schemes etc in order to improve life span of WSN. There also exist some problems in WSN:

- Path loss due to low bandwidth
- Loss of fusion accuracy
- More energy consumption

The data collectiona approachin WSN can be classified into multi path approaches, query propagation approaches, and mobile agent approaches. Multipath approaches achieve a high degree of reliability by making use of multiplepaths to send information from every sensor node to the collection point[15][16].

1.1 Mobile Agent concept

To meet the above challenges the concept of mobile agents has been proposed. To solve the problem of the overwhelming data traffic, bandwidth, Hairong et al, [1] proposed a mobile agent based distributed sensor network (MADSN) for scalable and energy-efficient data aggregation. Mobile Agents can be used in mobile computing environment for network control and management The Mobile Agent (MA) is a special kind of software which visits the network either periodically or on demand and performs data processing autonomously while migrating from node to node. There is very important property of mobile agents is their ability to autonomously move from one device to another (mobility). It is proved that mobile agent implementation can save up to 90 percents of data transfer time due to avoiding the raw data transfer, Feiyi Wang et al. [5]. Mobile agents are a distributed computing paradigm based on code mobility for high effectiveness and efficient in IP-based highly dynamic distributed environments A.R. Silva [2]. The two attributes SinkID and MA_SeqNum are used to identify an MA packet. Whenever the Sink sends a new MA packet, it increments the MA_SeqNum. The SrcList list
specifies the node itinerary LCF(Local Closest First) to be visited by the MA. NextSrc specifically determines the sequence of node identifiers that must be visited by the MA[7][8]. All mobile agents have the features: timer, navigation algorithm, network information, and above all stability of the network and load traffic of network.

In traditional network domain, data is collected by the individual sensor and is transferred to sink node for further processing known as data fusion as in figure 1.1. So in this type of network a large amount of data travels between the network elements. By using mobile agent data only on demand need to be sent to any sink node as in figure 1.2. Mobile agent is proposed to perform the following functions: (1) eliminating data redundancy among sensors by local processing at node level in the data context; (2) eliminating spatial redundancy among closely-located   sensors   by data aggregation at the task level; (3) reducing communication overhead by concatenating data at the combined task level.

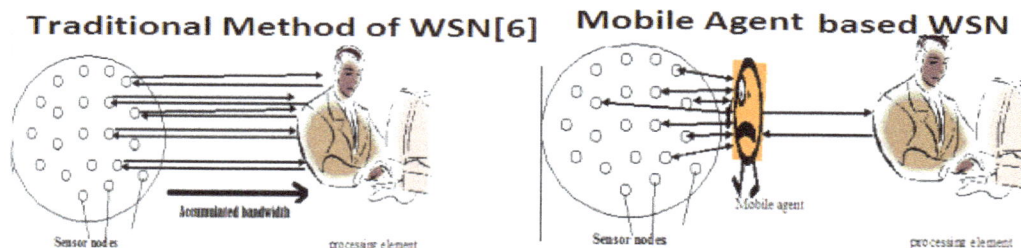

Fig. 1.1

Fig. 1.2

1.2 Work processing of Mobile Agents

The mobile agent to begin with when arrives at a node for the first time, stores its code at that node for future visit without carrying its code and sending the result to the sink node. Mobile agent also copies its processing code into the memory of each node in the first round. Once the whole task is completed, all those nodes discard the processing code. The agent performs the filtering function when it acquires the data. For example to estimate temperature, if node 1, node2 and node3 are neighbouring node, they must have similar temperature. We suppose that node1 estimate the temperature of thirty degree, node3 estimate the temperature of thirty two degree. node2 is located between node1 and node3. node2 must have the temperature either from thirty to thirty two or similar degree. But, if node2 estimates the temperature over forty or less than ten degree, Sensing data of nodes can be fault data. In this way agents can judge the data When the network connection is not available the mobile agent can save its status on secondary memory and later migrate to the home network when the connection is available. Software agents need not always travel across a network to communicate with information sources, or other agents. They work on message passing systems for agents.

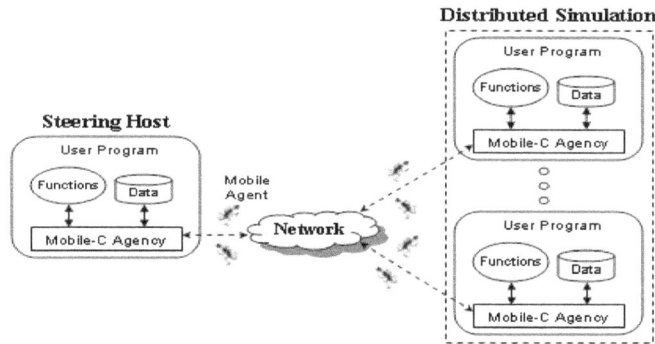

Figure 1.3 Multiple Criteria Clustering of Mobile Agents [9]

1.3 clustering of Mobile Agents

Cluster analysis is concerned with the grouping of alternatives into homogeneous clusters (groups) based on certain features. Three well-known clustering strategies are hierarchical clustering (Pandit, Srivastava, and Sharma 2001 ; Dias, Costa, and Climaco 1995), graph-theoretic methods (Matula 1977) and conceptual clustering (Michalski and Stepp 1982, 1983). Conventional clustering approaches involve two main factors: 1) distance (or dissimilarity) measurement, and 2) cluster centres. Head node has id information of nodes in its cluster. When the head node generates agents, the agent can have the id information of nodes that should pass. And the agent migrates to the nodes in pass route. And the agent also contains the information of neighboring nodes that are not included in the route. If node in route has the problems, the agent migrates to neighbouring node through the neighbouring information. By clustering mobile agents, cost of time for each agent to travel to its required routers, can be reduced. Routers have certain capabilities and can process certain requests of each intelligent agent.We propose in our present work a new application to clustering. We discuss how clustering methods can be applied and used for MCCMA communication network problems. Intelligent agents are grouped into cluster in such a way that each cluster has similarity in responding to a group of routers. In this way it helps to reduce demand imposed upon a network for a set of required task to be performed.

2. Related Work

Since sensor network works with limited energy,there has been an extensive research work toward energy efficiency of sensor. Liang Zhao [11] proposed a Medium-contention based Energy-efficient Distributed Clustering (MEDIC) scheme, through which sensors self-organize themselves into energy-efficient clusters by bidding for cluster headship. Sudhir [12] focuses on use of classification techniques using neural network to reduce the data traffic from the node and thereby reduce energy consumption. Low Energy Adaptive Clustering Hierarchy (LEACH) algorithm is applied to decompose the network into clusters, with one head for each. The main purpose of this technique is to collect the data by a mobile agent and to send them together to minimize the transmission . A number of papers have proposed algorithms for data Compression/Decompression (C/D) to reduce the amount of data transmitted by the sensors. The MA is a special kind of software that propagates over the network either periodically or on demand (when required by the applications). It performs data processing autonomously while migrating from node to node. Q. Wu et. al. [13]. In [14] *Costas Tsatsoulis et al* says that instead of one centralized and usually very large system that assumes the complete control and intelligence of the network, a number of smaller systems, or agents, can be used to help manage the network in a cooperative manner .This has motivated the multi agent systems (MAS) in telecommunication networks. The use of MAs in computer networks has certain advantages and disadvantages like code caching, safety and security, depending on the particular scenario. A lot of research has been done on power utilization in WSN various researchers have identified different clustering methods, routing algorithms, scheduling algorithms and load balancing algorithms.

In our proposed work, it is assumed that sensor network is divided into clusters. In traditional client/server approach, data is transmitted directly to sink node which will reduce the life span of sensor network. To overcome this approach clustering approach is used, in which sensors are clustered using clustering algorithm. Sensor nodes send all sensed data to sink node whether all data is necessary or not. This approach requires a lot of bandwidth and increase network traffic.

Mobile Agent is programmes which move sensor to sensor to gather information and transfer only specified information to sink node. This approach requires low bandwidth and reduce network traffic. Cost effective migration path is defined by the gateway as mobile agents visit the WSN. In this paper , Mobile Agents are divided into clusters based on multiple criteria. Then an incidence matrix is created and then agents are clustered using some rules defined below. The clustering of multiple criteria alternatives can also bring the following benefits:[9]

1) It decreases the set of alternatives - since the Decision Maker may be interested in only those alternatives with similar kinds of features and discard other alternatives.
2) It decreases the number of criteria - when evaluating alternatives of one group, the Decision Maker does not need to consider all criteria since one or more criteria values of the alternatives in the same group are equal or very close.
3) It provides a basis for more in-depth evaluation of alternatives - once one set of clustered alternatives are selected then this set can be explored in more depth for analysis, selection, and implementation purposes.
4) It may provide a basis for analyzing multiple criteria problems. Each decision maker may cluster alternatives differently, and hence, clustering of alternatives may provide a basis for negotiation.
5) In case of selection of a group of alternatives, each decision maker can be in charge of one clustered group;
Hence the designated decision maker selects the best solution from each clustered group *Miettnen and Salminen (1999) or Malakooti and Raman (2000)*.

3. Clustering for Mobile Agents

There are two hierarchical levels of agents: load management agents and parent agents. The load management agent visits every node in the network efficiently using Dijkstra's shortest path algorithm and it collects the necessary information to determine the optimal routes from all other nodes to that particular node. The parent agents control the next management level. They travel around the network and launch load agents where network management is needed. In this paper we use load management agents for collecting needed information form each and every node.

Each router has certain capabilities to process certain requests or needs of each intelligent agent. The problem is to group the intelligent mobile agents into a number of clusters such that each cluster has similarity in meeting router demands. Suppose that an intelligent mobile agent must visit a certain number of routers in order to complete the task that it is assigned to. If required, additional agents can be assigned of identical capabilities. Such acquiring will increase cost .Hence each agent can be duplicated. In a simpler way, the procedure is to find cluster center Cr and then cluster all alternative objects into a group according some rules. First of all, we define similarity measurement d(X,C) as the generalized Euclidean distance between objects Xi and center Cr , r=1,2,3,4……..n[7]

$$d\ (X_i, C_r)= \sqrt{k_1(x_1 - c_{11})^2 + k_2(x_2 - c_{12})^2 + \ldots + k_m(x_m - c_{1m})^2}$$

Where k1, k2……..k_m are the generalized Euclidean distance, the coefficients k = {k_i: i = 1, … m} can be used to represent indirectly the important index of each attribute of the alternatives. An objective X belongs to cluster r if and only if d(X, Cr) < d(X, Ct) l t = 1, 2... R and t < r.

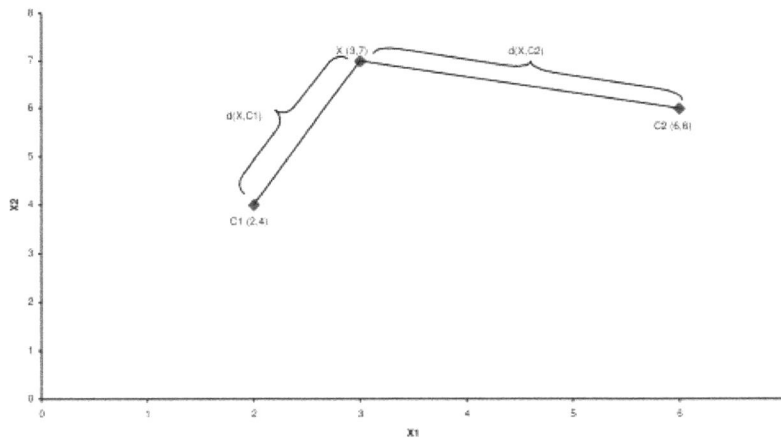

Figure 3.1 Example of Cluster Memberships

Clustering Algorithm steps:

Step 1: Consider unclassified wireless sensor network area of (X*Y) meter having N number of routers and M number of agents.

Step 2: Sink node dispatch mobile agents with requirement of specific task to wsn. Mobile agent migrate from first to last sensor node to perform specified task.

Step 3: Route table of agents is created corresponding to their routers.

Step 4:   A matrix {A_{ij}} called router-agent incidence matrix is created which has i rows of routers and j columns of agents. An element a_{ij} of matrix {A_{ij}} is 1 if agent j require operation to be perform on router i, otherwise a_{ij} is 0.

Step 5:   Based on route table and incidence matrix of router-agent family, agents are divided into clusters. Agents perform operations on some specific routers are put into clusters.

Step 6:   The few entries outside the diagonal blocks represent operations to be performed out-side the assigned group router cells. These elements are called exceptional elements.

We consider an example and divide into clusters

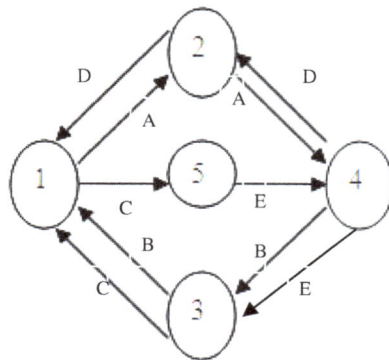

Figure 3.2 Path of routers

Agent	Routers to visit
A	1→2→4
B	4→3→1
C	3→1→5
D	4→2→1
E	5→4→3

Table 1 Agents path table

Cluster no.	Routers cells	Agents families
Cluster 1	1,5,3	B,C,E
Cluster 2	2,4	A,D,E

Table 2 (Agent family and visited Routers table)

Here Table 1 is divide the agents into two clusters. That is, the primary cluster for handling B,C,E agents corresponding router are 1,5,3 and for A,D,E agents corresponding routers are 2 and 4. Agents B,C,E perform some operation on router 1,5,3 and similarly A,D,E on router 2,4. Here in above example agent E used in both clusters so if we can afford the cost of more agents then we will acquire additional agents, the duplicated agents.

Agents \ Routers	A	B	C	D	E
1	1	1	1	1	0
2	1	0	0	1	0
3	0	1	1	0	1
4	1	0	0	1	1
5	0	0	1	0	1

Table 3 Incidence matrix of router-agent

The router-agent cell formation is a strategy to group routers into cells and agents into families. Families of agents can then be completely processed in their corresponding group router cells [10]. The processing of agents on routers can be represented in the form of a matrix called router-agent incidence matrix which has i rows of routers and j columns representing the agents.

4. Experimental work

The proposed work is shown in form of data flow diagram as follows:

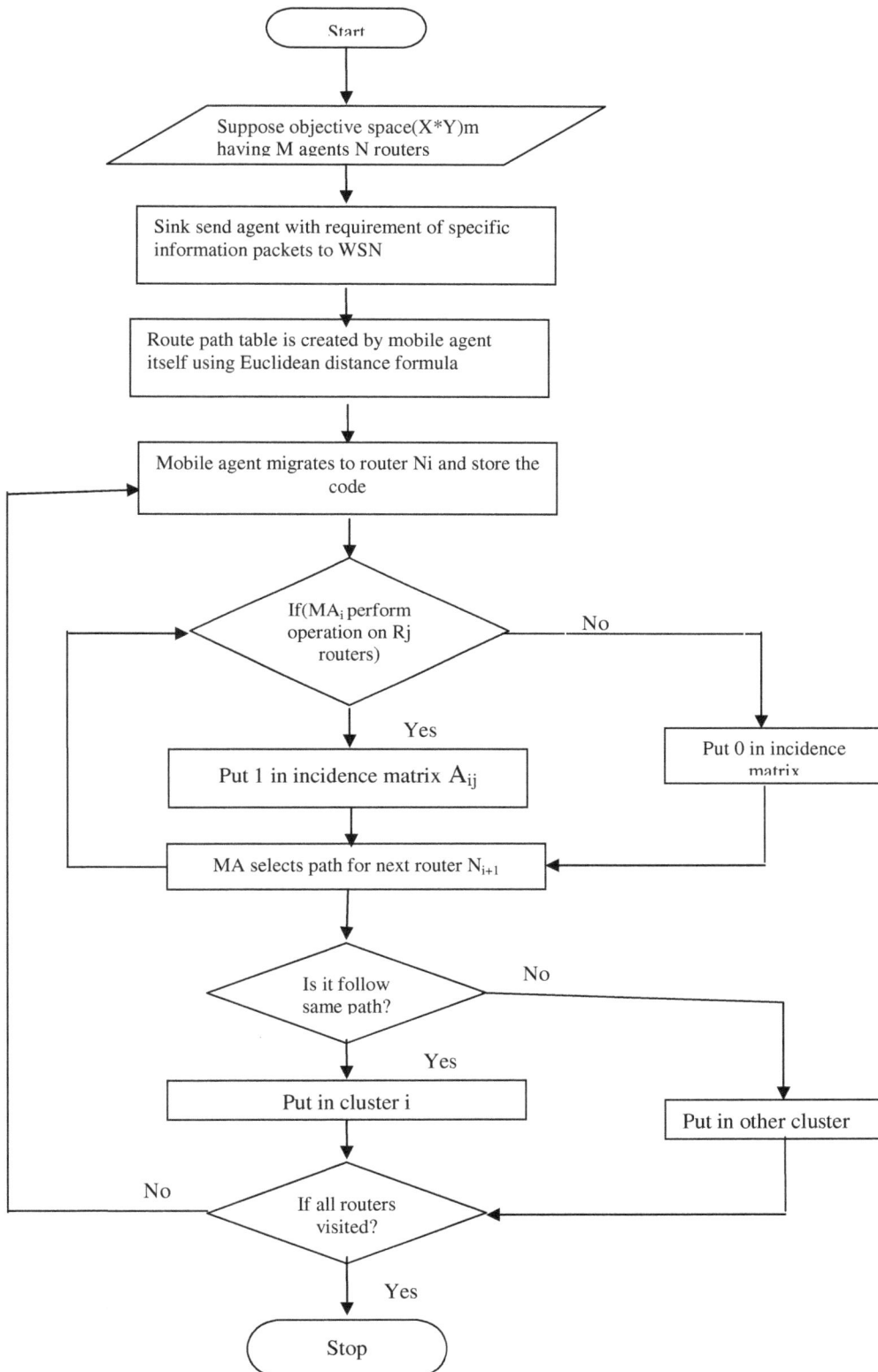

```
                        ┌──────────────┐
                        │    Start     │
                        └──────┬───────┘
                               │
                               ▼
                ╱────────────────────────────╱
               ╱ Suppose objective space(X*Y)m ╱
              ╱  having M agents N routers      ╱
             ╱────────────────────────────╱
                               │
                               ▼
                ┌────────────────────────────────┐
                │ Sink send agent with requirement│
                │ of specific information packets │
                │ to WSN                          │
                └────────────────┬───────────────┘
                                 │
                                 ▼
                ┌────────────────────────────────┐
                │ Route path table is created by  │
                │ mobile agent itself using       │
                │ Euclidean distance formula      │
                └────────────────┬───────────────┘
                                 │
                                 ▼
                ┌────────────────────────────────┐
                │ Mobile agent migrates to router │
                │ Ni and store the code           │
                └────────────────┬───────────────┘
                                 │
                                 ▼
                         ◇ If(MAi perform ◇── No ──┐
                         ◇ operation on Rj◇         │
                         ◇ routers)       ◇         │
                             │ Yes                  ▼
                             ▼              ┌──────────────┐
                ┌──────────────────────┐   │ Put 0 in     │
                │ Put 1 in incidence   │   │ incidence    │
                │ matrix Aij           │   │ matrix       │
                └──────────┬───────────┘   └──────┬───────┘
                           │                      │
                           ▼                      │
                ┌──────────────────────┐◄─────────┘
                │ MA selects path for   │
                │ next router Ni+1      │
                └──────────┬───────────┘
                           │
                           ▼
                      ◇ Is it follow ◇── No ──┐
                      ◇ same path?   ◇         │
                           │ Yes              │
                           ▼                  ▼
                ┌──────────────────┐   ┌──────────────┐
                │ Put in cluster i │   │ Put in other │
                └────────┬─────────┘   │ cluster      │
                         │             └──────┬───────┘
                         ▼                    │
          ┌── No ── ◇ If all routers ◇◄───────┘
          │         ◇ visited?       ◇
          │              │ Yes
          │              ▼
          │         ┌──────────┐
          │         │   Stop   │
          │         └──────────┘
```

$$ \text{If(MA}_i \text{ perform operation on R}_j \text{ routers)} $$

$$ \text{Put 1 in incidence matrix } A_{ij} $$

$$ \text{MA selects path for next router } N_{i+1} $$

Now we take an example of 15*21 incidence matrix having 15routers and 21 agents grouping agents into families and routers into cells. Then result show block diagonal matrix which is clusters of agents based on some criteria stated above.

Agent / Routers	1	2	3	4	5	6	7	8	9	10	11	12	13	14	15	16	17	18	19	20	21
1		1				1	1				1					1	1		1	1	
2				1				1	1			1		1				1			
3	1			1			1						1					1			
4			1		1								1					1			
5								1	1	1			1		1				1		
6	1						1				1	1								1	
7		1					1					1								1	
8	1		1	1		1												1			
9	1				1	1								1				1			
10								1	1	1					1						
11		1					1				1	1								1	
12																1	1		1		
13				1				1	1	1			1		1						
14		1															1		1		1
15			1				1									1	1		1		1

Table 5 Router-Agent initially incidence matrix

Now we divide above incidence matrix of agents and router according agent families and router cells as follows:

Routers   **Agents** ⟶

Routers	2	7	11	12	20	1	3	4	6	14	18	5	8	9	10	13	15	16	17	19	21
1	1	1	1	0	1															1	
6	0	1	1	1	1																
7	1	1	0	1	1																
11	1	1	1	1	1																
3						1	0	1	0	1	1										
4						0	1	0	1	1	1										
8						1	1	1	1	0	1										
9						1	0	0	1	1	1										
2												1	0	1	1	1	1				
5												0	1	1	1	1	1		1		
10												1	1	1	1	0	1				
13												1	1	1	1	1	1				
12																		1	1	1	0
14									1									0	1	1	1
15	1								1									1	1	1	1

Table 7 Clustered Matrix

Cluster no.	Routers cells	Agents families
1	1,6,7,11	2,7,11,12,20
2	2,5,10,13	5,8,9,10,13,15
3	3,4,8,9	1,3,4,6,14,18
4	12,14,15	16,17,19,21

Table 6(Agent family and visited Routers table)

The resultant matrix after grouping (according above table) has four distinct router-agent cells. Table 6 shows that agent family 1, consisting of agents 2, 7, 11, 12 and 20 can be processed in group router cell 1,which contains routers 1, 6, and 7 and 11 in similar ways family 2 ,consisting agents 1,3,4,6,14,18 process on routers 3,4,9 and in same way family3, 4. Here in table it is show that agents are clustered corresponding to their routers. The few entries outside the diagonal blocks represent operations to be performed outside the assigned group router cells.   These elements are called exceptional elements. The corresponding router is called a bottleneck router, and the corresponding agent is called an exceptional agent.

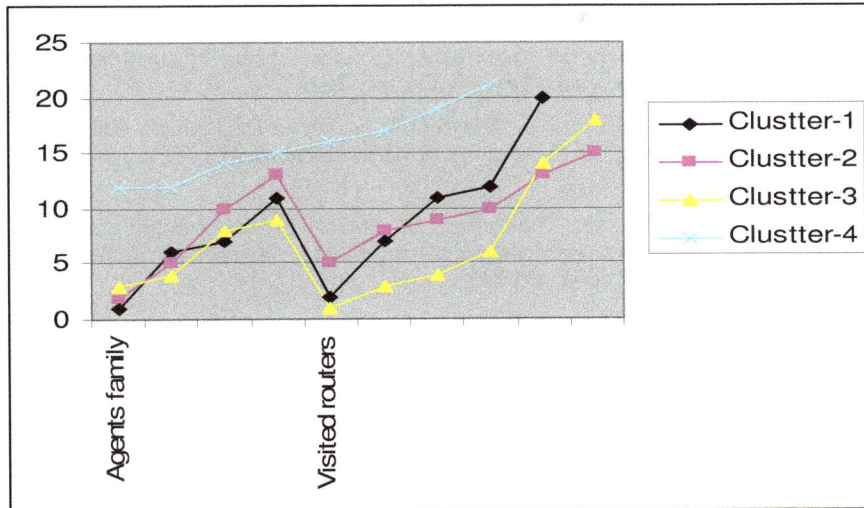

Figure 4.1 Clustering of router-Agent family

5. Conclusion

 In this paper, we discuss how clusters can be selected and how different alternatives could be clustered into independent clustered groups. Here our experimented result is shown in form of graph of router-agent family. Incidence matrix is divided into four clusters. In this paper, some fundamental definitions and algorithms are formulated for clustering intelligent mobile agents. We develop a method for Multiple Criteria Clustering of Mobile Agents (MCCMA) for wireless sensor network systems. Intelligent mobile agents were introduced as agents those are sent to routers to carry out a set of required tasks. When clustering intelligent mobile agents with similar functional capabilities, the purpose is to minimize cost or travel time of each agent visiting its required routers. It will decrease network load for a set of required tasks to be performed. We demonstrate how intelligent agents are clustered into agent family and router cells by making a number of comparisons. The future development of this work includes clustering physical objects presented graphically to the decision maker; applying the selection of the most preferred alternative for each cluster, and considering the problem of multiple decision makers.

REFERENCES

[1]   Hairong Qi, Yingyue Xu, Xiaoling Wang, "Mobile-Agent Based Collaborative Signal and Information Processing in Sensor Networks," in Proceeding of the IEEE, Vol. 91, NO. 8, pp.1172-1183,Aug.2003

[2]   A.R. Silva, A. Romao, D. Deugo, M. Mira da Silva, "Towards a Reference Model for Surveying Mobile Agent Systems," Autonomous Agent and Multi-Agent Systems, Vol. 4, N. 3, pp.187–231, 2001.

13]   M.Bendjima,M.Feham, "Intelligent Wireless Sensor Network Management Based on a Multi-Agent System" in   proceeding of International Journal of Computer Science and Telecommunications, Vol. 3, ISSN 2047-3338, Issue 2, February2012

[4]   Bernhard Bösch , Rodrigo S. Allgayer "Mobile Agents Model and Performance Analysis of a Wireless Sensor Network Target Tracking Application" in Springer-Verlag Berlin Heidelberg, LNCS 6869, pp. 274–286, 2011

[5]   Hairong Qi and   Feiyi Wang "Optimal Itinerary Analysis for Mobile Agents in Ad hoc Wireless Sensor   Networks", published   in   the   13th   international conference on wireless communication Canada, Pp- 147- 152 July 2001.

[6]   Ankit Jagga, Kuldeep, Vijay Rana, "A Hybrid Approach for Deploying Mobile Agents in Wireless Sensor Network", in Proceedings of International Journal of Computer Applications® (IJCA), 2012.

[7]   Malakooti, B., V. Raman, "Clustering and Selection of Multiple Criteria Alternatives using Unsupervised and Supervised Neural Networks", Journal of Intelligent Manufacturing , Vol. 11, 2000, 435-453.

[8]   Miettnen, K. and P. Salminen, "Decision-Aid for Discrete Multiple Criteria Decision Making Problems with Imprecise Data," European Journal of Operational Research, 119, 50-60. 1999

[9]
        www.google.co.in/search?tbm=isch&hl=en&source=hp&biw=1024&bih=566&q=cluster+of+mobile+

[10]   Chan, H.M. Milner, D.A., "Direct Clustering Algorithm for Group Formation in Cellular Manufacturing," Journal of Manufacturing Systems, 65-75, 1982

[11]   Liang Zhao, Qilian Liang, "Medium-contention based energy-efficient distributed clustering (MEDIC) for wireless sensor networks", International Journal of Distributed Sensor Networks, 3, 2007, 347-369

[12]   Sudhir G. Akojwar, Rajendra M. Patrikar," Improving life time of wireless sensor networks using neural network based classification techniques with cooperative routing", International Journal of Communications, vol. 2, no. 1, 2008

[13]   Wu, Q., Rao, N.S.V., Barhen, J., etc, "On computing mobile agent routes for data fusion in distributed sensor networks," IEEE Transactions on Knowledge and Data Engineering, Vol.16 , NO. 6,   pp. 740-753, June 2004

[14]   Costas Tsatsoulis and Leen-Kiat Soh, "Intelligent Agents in Telecommunication Networks", Computational Intelligence in Telecommunications Networks, W. Pedrycz and A.V. Vasilakos (Eds.), CRC Press, 2000

[15]   B. Deb, S. Bhatnagar, and B. Nath, "Reinform: Reliable information forwarding using multiple paths in sensor networks," in IEEE International Conference on Local Computer Networks (LCN'03), 2003.

[16]   D. Ganesan, R. Govindan, S. Shenker, and D. Estrin, "Highly-resilient, energy-efficient multipath routing in wireless sensor networks," ACM Mobile Computing and Communications Review, 2003.

[17]   *Kazem Sohraby, Daniel Minoli, Taieb Znati, A John Willy & Sons,"Wireless Sensor Networks Technology, Protocols, and Applications" Inc Publications, 2007.*

[18]   Malakooti, B., "A Decision Support System and a Heuristic Interactive Approach for Solving Discrete Multiple Criteria Problems," IEEE Trans. Syst. Man Cybern., 18, 273-285. 1988.

[19]   Malakooti, B. "Ranking and Screening Multiple Criteria Alternatives with Partial Information and use of Ordinal and Cardinal Strength of Preferences", IEEE Trans. on Systems, Man, and Cybernetics Part A, Vol. 30, 3, 355-369, 2000.

[20]   Zeleny, M., Linear Multiobjective Programming, New York: Springer-Verlag, 1974.

[21]   http://en.wikipedia.org/wiki/Software_agent#Definition

8

Load Balancing with Reduced Unnecessary Handoff in Energy Efficient Macro/Femto-cell based BWA Networks

Prasun Chowdhury[1], Anindita Kundu[2], Iti Saha Misra[3], Salil K Sanyal[4]

Department of Electronics and Telecommunication Engineering, Jadavpur University, Kolkata-700032, India
[1]prasun.jucal@gmail.com, [2]kundu.anindita@gmail.com,
[3]itisahamisra@yahoo.co.in, [4]s_sanyal@ieee.org

ABSTRACT

The hierarchical macro/femto cell based BWA networks are observed to be quite promising for mobile operators as it improves their network coverage and capacity at the outskirt of the macro cell. However, this new technology introduces increased number of macro/femto handoff and wastage of electrical energy which in turn may affect the system performance. Users moving with high velocity or undergoing real-time transmission suffers degraded performance due to huge number of unnecessary macro/femto handoff. On the other hand, huge amount of electrical energy is wasted when a femto BS is active in the network but remains unutilized due to low network load. Our proposed energy efficient handoff decision algorithm eliminates the unnecessary handoff while balancing the load of the macro and femto cells at minimal energy consumption. The performance of the proposed algorithm is analyzed using Continuous Time Markov Chain (CTMC) Model. In addition, we have also contributed a method to determine the balanced threshold level of the received signal strength (RSS) from macro base station (BS). The balanced threshold level provides equal load distribution of the mobile users to the macro and femto BSs. The balanced threshold level is evaluated based on the distant location of the femto cells for small scaled networks. Numerical analysis shows that threshold level above the balanced threshold results in higher load distribution of the mobile users to the femto BSs.

KEYWORDS

Hierarchical BWA Networks, Handoff; Continuous Time Markov Chain, QoS Management, Load Balancing, Energy Efficient Femto BS

1. INTRODUCTION

The recent development of hierarchical macro/femto cell based broadband wireless access (BWA) networks is drawing the attention of wireless service providers more than ever due to their enhanced indoor coverage. The hierarchical network architecture not only provides extension in the cell coverage but also provides increase in the capacity and service quality enhancement. In BWA networks like WiMAX [1], femtocells are cost effective means to provide ubiquitious connectivity. The femto cellular base station is a miniaturized low-cost and low-power Base Station (BS) which uses a general broadband access network as its backhaul [2]. With the introduction of femto cells the total number of active users in the service area increases due to capacity enhancement. However, the mobility of these active users leads to huge number of macro/femto handoff in the hierarchical cell structures. On the other hand, if the load of the network is low most of the femto BSs remain unutilized even though it consumes power. Hence the power conservation of the entire hierarchical network along with elimination of unnecessary handoff provides a significant area of research work.

In recent literatures like [3], [4] authors have presented WiMAX femto cell system architectures and evaluate its performance in terms of network coverage, system capacity and performance of mobile station in indoor environment. On the other hand, authors of [5], [6] have compared the performance in private and public access method in WiMAX femto cell environment. However, the process of handoff, QoS requirement of the mobile stations and reducing the energy consumption of the femto BSs have not been considered in any of the aforementioned papers.

A variety of handoff algorithms based on received signal strength (RSS) have been considered in [7], [8], [9]. Velocity of the mobile node have been considered in [8], [9] as a parameter for handoff decision. However, QoS guarantee for the real time service and conservation of energy of the hierarchical network has not been considered in these papers too.

In [10] authors have proposed a new handoff algorithm to correctly assign the users to the femto cells but QoS profile, energy conservation and network load balancing are not taken into account. Moreover, no relation between the femto cells and the whole system has been considered in their simulation which does not reflect a real mobile WiMAX architecture.

In this paper, we have considered a WiMAX system where a mobile station is moving from a macro cell to a femto cell. Since at this stage the distance of the mobile station from the macro BS is more than the femto BS, more power will be consumed by the mobile station to communicate with the macro BS than the femto BS. Hence, we have assumed that the mobile station gives higher priority to the femto BSs than the macro BS. Thus a mobile staion selects the femto BS as its serving BS when it receives siganal from both the macro and femto BSs as well as the RSS from macro BS falls below its threshold level. In this paper, we have also determined the balanced threshold level of RSS from macro BS based on the distant location of the femto cells for small scaled network. Balanced threshold level provides equal load distribution of the mobile users to the macro and femto BSs. Numerical analysis shows that threshold level above the balanced threshold results in higher load distribution of the mobile users to the femto BSs.

To achieve QoS aware hierarchical networks, our handoff decision algorithm is based on two main factors viz. the velocity of mobile station and the service type of the mobile station. When a user moves with an ongoing call at a very high velocity (above velocity threshold) from one end of the hierarchical cell to the other end, it is expected that the user will experience huge number of macro/femto handoff in a short period of time. This burdens the overhead of the macro BS. A considerable amount of packet loss may also be encountered which degrades the call quality. On the other hand, though a user moves with a velocity lower than the velocity threshold it experiences comparitively lower handoff rate but handoff still happens. In this case, if the ongoing call is a real-time service then packet loss will hamper the call quality.

To avoid this quality degradation due to multiple number of macro/femto handoff, in our handoff decision algorithm we have considered a velocity threshold such that any user moving with a velocity higher than the velocity threshold or undergoing real-time transmission will not undergo any macro/femto handoff. Thereby, we eliminate unnecessary handoff and provide reduced network overhead and increased user satisfaction.

However, with the introduction of the femto cells the power consumption of the entire network increases. The femto BSs may consume power even if no end user resides under its coverage. This may lead to wastge of power. To overcum this wastage of power we have considered a power conservation scheme by which the femto BSs will remain in a low power 'Idle mode' [11] with the pilot transmissions and processing switched off when no user is present in its coverage. Based on this considerartion our proposed handoff decision algorithm has been analyzed using Continuous Time Markov Chain (CTMC) Model.

The remainder of the paper is organized as follows: Section 2 discusses system model and the detailed description of the proposed energy efficient handoff decision algorithm of WiMAX macro/femto-cell networks. Section 3 shows analytical model and QoS performance evaluation parameters. Numerical results are discussed in section 4. Finally, Section 5 concludes the paper.

2. SYSTEM MODEL AND PROPOSED ENERGY EFFICIENT HANDOFF DECISION ALGORITHM

A WiMAX macro cell of 1.2 km radius is considered along with multiple femto cells deployed randomly at a distance of atleast 'R' meter from the macro BS as shown in Figure 1.

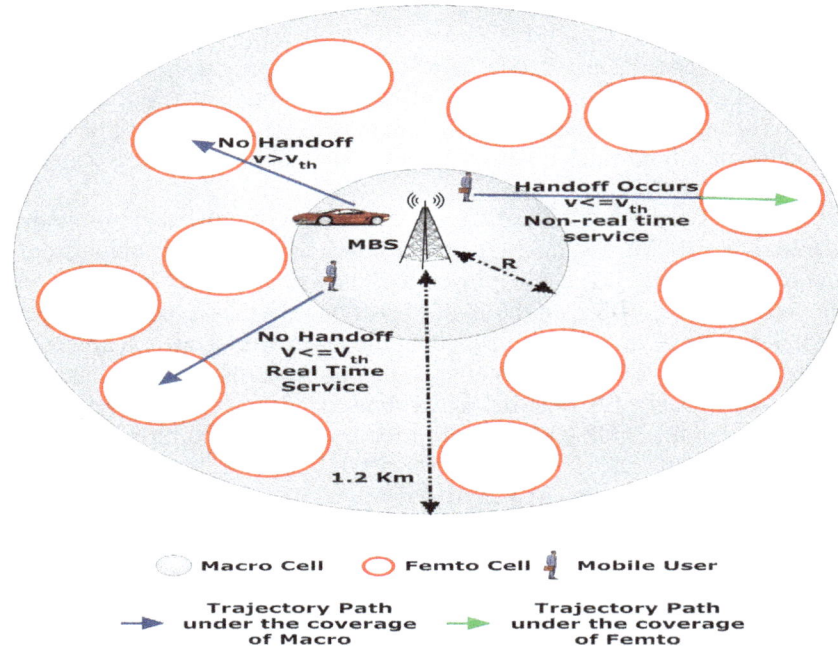

Figure 1. Energy efficient handoff decisions in the hierarchical system model

No femto cell is considered within 'R' meter of radius of the macro BS because the RSS of the mobile nodes residing within this area is assumed to be quite high. A number of mobile users are deployed randomly under the coverage of the macro BS with varied velocity and undergoing calls of varied service type. The rest of the system model parameters are shown in Table 1.

TABLE 1. SYSTEM MODEL PARAMETERS

Parameters	Value
Macro cell radius	1.2 Km
Femto cell radius	30m
Real-time service type	UGS, rtPS
Non real-time service type	nrtPS, BE

Initially, the femto BSs are considered to be in the idle mode with all the pilot transmissions and associated radio processing disabled. The femto BSs incudes a low power sniffer (P_{sniff}) [11] which allows the detection of an active call originating from a mobile under its coverage to the macro BS. At this stage, the femto BS changes to active mode and requests the macro BS to handoff the newly originated call to it. Thus, the femto BSs are active only when any end user is active under its coverage which thereby enhancing the energy conservation.

Macro/femto cellular handoff comprises of two main phases – handoff strategy and handoff decision algorithm. The first phase deals with the RSS comparison while in the second phase the system decides when to trigger the handoff. In this paper, we contribute an efficent algorithm for the second phase. The handoff strategy proposed in [7] has been considered for the first phase.

Let RSS_m and RSS_f denote the received signal strength from the macro BS and femto BS experienced by a mobile node at any instant of time. As a mobile node moves in a straight line from the macro BS to femto BS with constant low velocity as shown in Figure 1, conventional handoff occurs. The conventional handoff algorithm with the RSS comaparison [7] can be expressed as in equation (1).

$$RSS_m < RSS_{m,th} \quad \text{and} \quad RSS_f > RSS_m + \Delta \tag{1}$$

where $RSS_{m,th}$ and Δ denotes the minimum RSS threshold level from the macro BS and the value of hysteresis respectively.

The pathloss encountered by a mobile node as it moves away from the macro BS diminishes the RSS_m. As the distance from the macro BS increases, this pathloss triggers the handoff situation where RSS_f becomes higher than RSS_m. In our scenario, we have considered the ITU pathloss model in slow fading channel [12] as shown in equations (2) and (3).

$$PL_m = 15.3 + 37.6\log_{10}(D) + PL_{hw} \quad \text{where,} \quad PL_{hw} = 10 \tag{2}$$

$$PL_f = 38.46 + 20\log_{10}(d) + 0.7d \tag{3}$$

where PL_m and PL_f denotes the pathloss from macro and femto BS respectively. 'D' and 'd' are the corresponding distance of the mobile user from macro and femto BS.

Thus the resulting RSS_m and RSS_f encountered by the mobile node are shown in equations (4) and (5) respectively.

$$RSS_m = P_{m,tx} - PL_m \tag{4}$$

$$RSS_f = P_{f,tx} - PL_f \tag{5}$$

where $P_{m,tx}$ and $P_{f,tx}$ denote the transmit power of macro BS and femto BS respectively.

However, in our proposed handoff decision phase the conventional handoff is not adopted in the following two cases.

Case I: When the mobile user is moving at a very high velocity

Conventional handoff is not applicable when a mobile user is moving with a very high velocity. As a user moves with a very high velocity it undergoes huge number of macro/femto handoff within a very short period of time. The overhead of the macro BS thus increases unnecessarily.

Hence, in this paper we have considered a velocity threshold 'V$_{th}$' of 10 Kmph (non motor vehicular speed) and simulated our scenario accordingly. If a user moves with a velocity 'V' such that V> V$_{th}$, unlike conventional scenario the user will not undergo handoff. Thus the unnecessary handoff is eliminated and improved QoS is guaranteed.

Case II: When the mobile user is undergoing a UGS or rtPS i.e. real-time call

When a user is moving with a real-time connection the number of handoff encountered degrades the call quality proportionately. Hence, in our scenario, no handoff is triggered for them in order to maintain the call quality. Thereby unnesecessary handoff count decreases and improved QoS is assured to the real-time users.

In our scenario, the QoS guarantee achieved by considering only case I is refered to as soft QoS guarantee while QoS guarantee achieved by considering both case I and case II is called hard QoS guarantee. Soft QoS guarantee only will reduce the overhead of the network. On the other hand, hard QoS guarantee will reduce the network overhead as well as increase user satisfaction.

3. ANALYTICAL MODEL AND PERFORMANCE EVALUATION PARAMETERS

In this paper, the performance evaluation of the WiMAX macro/femto-cell networks is obtained by using Continuous Time Markov Chain (CTMC) Model [13]. In addition, we have considered Pareto distribution for the arrival process of the priority service types, so the network model experiences a flow of service requests in a continuous time domain. The network undergoes a continuous change in its current state due to the occurrence of events (i.e. arrival and service of priority calls). It is necessary to observe the short-lived states of the network in order to analyze its performance more accurately. This is only possible if the network is modelled with CTMC.

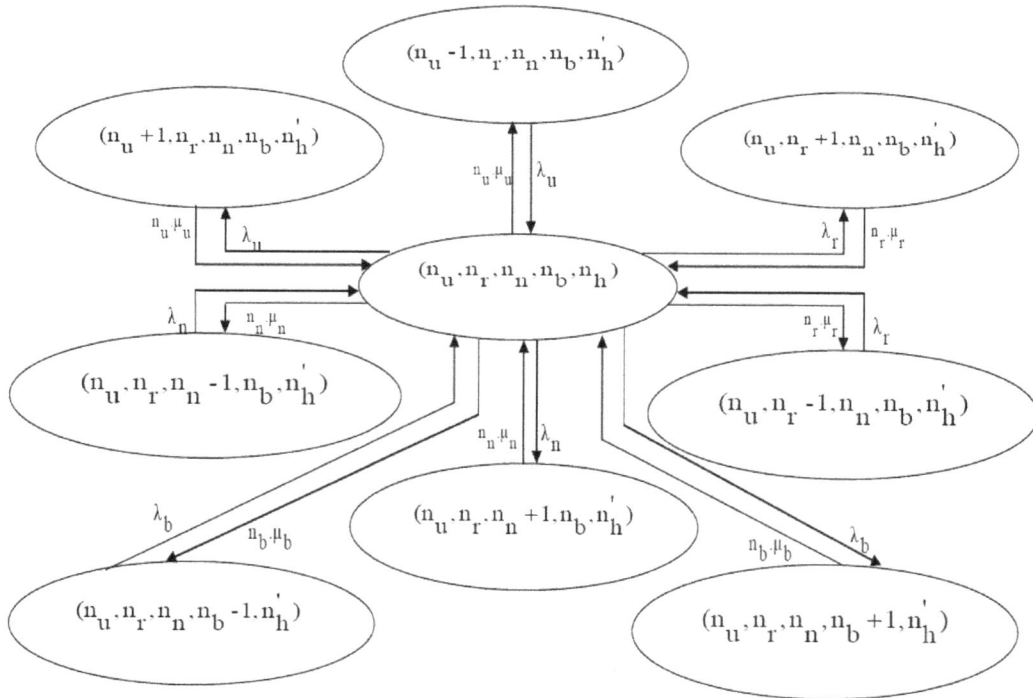

Figure 2. State transition diagram of the hierarchical WiMAX networks

A hierarchical WiMAX networks consisting of single macro BS along with multiple femto BS is considered. The macro BS will receive the handoff requests from the users directly. Four types of services i.e. UGS, rtPS, nrtPS and BE need QoS guarantees and request for a handoff whenever it finds any suitable femto BS in the near vicinity. The hierarchical networks change state from one to another upon the admission or termination of a service type. Further, it is assumed that the hierarchical networks either admit or terminate only one service type at a particular instance of time. So the next state of the hierarchical networks depends only on the present state of the hierarchical networks but does not depend on the previous states of the hierarchical networks. Therefore, the states of the hierarchical networks form a Markov Chain and accordingly the hierarchical networks can be analytically modelled as shown in Figure 2. In this scenario the hierarchical networks can uniquely be represented in the form of a five dimensional Markov Chain $(n_u, n_r, n_n, n_b, n_h)$ based on the number of services residing within the hierarchical networks and the total number of macro/femto handoff occurred in the network.

State s = $(n_u, n_r, n_n, n_b, n_h)$ represents that the hierarchical networks have currently admitted 'n_u', 'n_r', 'n_n' and 'n_b' number of UGS, rtPS, nrtPS and BE service respectively. 'n_h' represents the total number of macro/femto handoff occurred in that state of the hierarchical networks. We have assumed that initially no users are present under the coverage of femto cells. Hence, the total number of users present under the coverage area of femto BSs is also indicated by the parameter 'n_h'. In Figure 2, n_h' is the modified values of the variable 'n_h' after state transition. Pareto distribution is considered for the arrival process of the newly originated UGS, rtPS, nrtPS and BE with rates of $\lambda_u, \lambda_r, \lambda_n$ and λ_b respectively. This is because Pareto distribution supports more practical traffic model [14]. However, Poisson distribution is an ideal model, which is not practical in real WiMAX networks. The service times of UGS, rtPS, nrtPS and BE connections are exponentially distributed with mean $1/\mu_u, 1/\mu_r, 1/\mu_n$ and $1/\mu_b$ respectively.

Let the steady state probability of the state s = $(n_u, n_r, n_n, n_b, n_h)$ be represented by $\pi_{(n_u, n_r, n_n, n_b, n_h)}(s)$. As the Markov chain is irreducible, thereby observing the outgoing and incoming states for a given state 's', the steady state probabilities of all states of the hierarchical networks have been evaluated.

From the steady state probabilities we can determine various QoS performance parameters of the system as given below.

3.1. Handoff Probability (HO_Prob)

Number of handoff occurred in a particular state of the hierarchical networks multiplied with the steady state probability of that state will give the handoff probability of that particular state. Thereby, handoff probability of the hierarchical networks is obtained by summing the handoff probabilities of all the states of the hierarchical networks.

Hence, the probability of handoff occurred in hierarchical macro/femto networks can be calculated as follows.

$$HO\_Prob = \sum_{\forall s} n_h * \pi_{(n_u, n_r, n_n, n_b, n_h)}(s)$$

(6)

3.2. Macro Load (ML)

The Macro load is defined as the ratio of the number of users residing under macro BS to the total number of users present within the macro/femto hierarchical networks. ML can be calculated as follows.

$$ML = \sum_{\forall s} \frac{(n_u + n_r + n_n + n_b - n_h) * \pi_{(n_u,n_r,n_n,n_b,n_h)}(s)}{n_u + n_r + n_n + n_b} \tag{7}$$

3.3. Femto Load (FL)

The Femto load is defined as the ratio of the number of handoff users residing under femto BSs to the total number of users present within the macro/femto hierarchical networks. FL can be calculated as follows.

$$FL = \sum_{\forall s} \frac{n_h * \pi_{(n_u,n_r,n_n,n_b,n_h)}(s)}{n_u + n_r + n_n + n_b} \tag{8}$$

3.4. Energy consumption with active-idle mode ($E_{active\text{-}idle}$)

The probability that a femto BS is in active state is directly proportional to the network load and macro to femto handoff probability. In our markov chain model as we have considered exponential distribution of the service time of the users, hence the probability that a femto BS is in active state can be calculated as follows

$$\Pr ob(active) = (1 - e^{-\rho}) * HO\_prob \tag{9}$$

Where $\rho = \lambda / \mu$ for a particular service type and is termed as network load [15].

With the above consideration, probability that a given femto BS is in idle state is given as follows

$$\Pr ob(idle) = 1 - \Pr ob(active) \tag{10}$$

The power consumption in active state and idle state are denoted as 'P_{active}' and 'P_{idle}' respectively. The power of 'P_{sniff}' is also additionally consumed in idle state. Hence, monthly energy consumption of a single femto BS in kWh with active-idle mode is given as follows

$$E_{active-idle} = (P_{active} * \Pr ob(active) + (P_{idle} + P_{sniff}) * prob(idle)) * 3600 * 24 * 30 \tag{11}$$

3.5. Energy consumption with conventional mode ($E_{conventional}$)

Since femto BSs are always remain in active state in the conventional mode so in this case the monthly energy consumption in kWh is given as follows

$$E_{convention\ al} = P_{active} * 3600 * 24 * 30 \tag{12}$$

4. NUMERICAL RESULTS AND DISCUSSIONS

The contribution of this paper lies in balancing the network load between the macro BS and femto BSs and the reduction of the unnecessary handoff at minimal energy consumption. Exhaustive simulations have been carried out under MATLAB version 7.3. Since our main aim is to reduce unnecessary handoff not the handoff latency, in our simulation the value of hysteresis has been taken as zero. The arrival rates of all the connections are assumed to be same i.e. $\lambda_u = \lambda_r = \lambda_n = \lambda_b$. The values of the rest of the simulation parameters are shown in Table 2. In Table 2 the macro and femto transmit power has been taken from [16]. In ideal case scenario, it is assumed that there is no instrumental power loss in the femto cells and hence we have considered that transmit power is equal to their input active power i.e. 20dBm=100mW. The results associated with the load balancing are shown in Figure 3, 4 and 5 while reduction of the unnecessary handoff is exhibited in Figure 6. Also, the monthly energy consumption for a single femto BS with the proposed handoff reduction technique is given in Figure 7, 8 and 9. Justifications behind all the numerical results have also been provided.

TABLE 2. SIMULATION PARAMETERS

Parameters	Value
Macro transmit power ($P_{m,tx}$)	46dBm
Femto transmit power ($P_{f,tx}$)	20dBm
Femto active power (P_{active})	100mW
Femto idle power (P_{idle})	60mW
Low power sniffer (P_{sniff})	3mW
Velocity threshold (V_{th})	10kmph
$\mu_u = \mu_r = \mu_n = \mu_b$	0.2
Traffic ratio of UGS, rtPS, nrtPS, BE	1:1:1:1
Femto cell deployment	Random

As the threshold level of RSS_m or $RSS_{m,th}$ increases, the load of the macro BS decreases and the load of the femto BSs increases. This is revealed from Figure 3a and 3b respectively. With increase in the value of $RSS_{m,th}$ the macro/femto handoff count increases. Hence, the macro load decreases while increasing the femto BSs load.

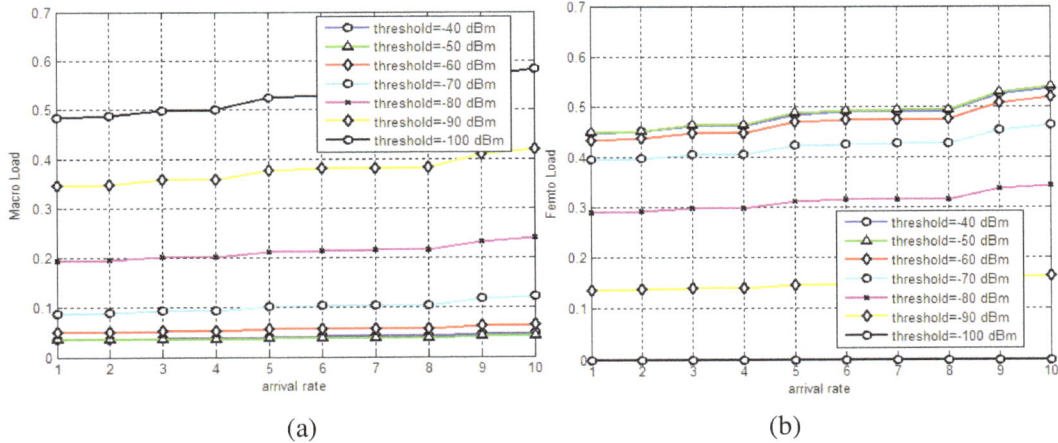

Figure 3. (a) Macro Load for various $RSS_{m,th}$ when 'R'=100m and (b) Femto Load for various $RSS_{m,th}$ when 'R'=100m

A saturation is observed in the load of both macro and femto BSs when $RSS_{m,th}$ reaches -50 dBm. No change is observed when value of $RSS_{m,th}$ increased further. It happens due to absence of femto cells within 'R'=100 m radius of macro BS as shown in Figure 1. Hence, when $RSS_{m,th}$ goes above -50 dBm, the mobile nodes residing within 'R'=100 m of macro BS do not find any femto BS. So no handoff is triggered and the load remains unchanged. Again, femto load is found to be zero at $RSS_{m,th}$= -100dBm due to macro cell outage. At this level of RSS theshold no femto cells are present to trigger handoff.

To balance the load of the macro BS and femto BSs, a study of the variation of their load has been performed with respect to the $RSS_{m,th}$. This variation is performed when femto cells are located at 'R'=100m apart from macro BS and is shown in Figure 4. The point of intersection of these two variations provides the value of $RSS_{m,th}$ at which the macro and femto load are observed to be same. It is observed that at $RSS_{m,th}$ =-83.4 dBm a balance between the load of the macro BS and femto BSs is achieved. Henceforth, Macro threshold level ($RSS_{m,th}$) at which load balancing is achieved is referred to as balanced threshold level. Load distribution to the femto BSs increases as the $RSS_{m,th}$ goes above the balanced threshold level i.e. -83.4 dBm.

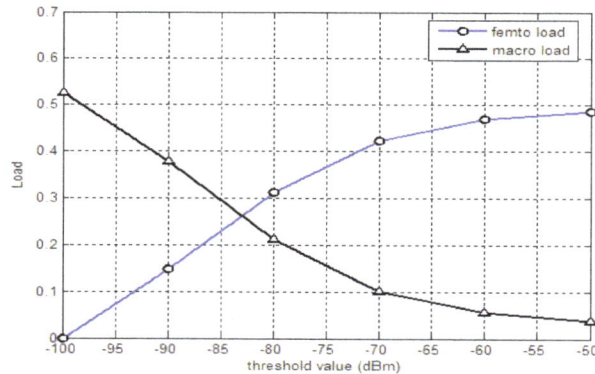

Figure 4. Comparison of macro/femto load for various $RSS_{m,th}$ when 'R'=100m

The above mentioned scenario has been generalized by varying the parameter 'R'. The corresponding balanced threshold level is evaluated in the similar way and is shown in Table 3. Thus, in order to have higher load distribution to the femto cells the macro BS should set the macro threshold level above the balanced threshold level with respect to the value of 'R'. The method of determining the balanced threshold level has been assessed for small scaled networks as the simulations are very time-consuming for broad scaled networks. This method can also be applied for broad scaled networks conceptually to calculate the corresponding balanced threshold level.

TABLE 3. THRESHOLD LEVEL FOR LOAD BALANCING

Variation of 'R' (meter)	Balanced threshold level (dBm)
200	-85.7
300	-87.1
400	-88.6
500	-89.8
600	-90.9
700	-92.1
800	-93.8
900	-95
1000	-95
1100	-95

In Table 3, as the value of 'R' increases, the balanced threshold level is observed to decrease gradually unless 'R' reaches 900 meters. From 900 meters onwards the macro threshold level is observed to remain constant at -95 dBm. The reason behind this is elaborated in Figure 5. Figure 5 shows the variation of the RSS of the mobile station with respect to its distance 'D' from the macro BS as obtained from equation (4). From Figure 5 we see that as the mobile station reaches the outskirt of the macro cell edge i.e. when D= 1.2 km, the RSS encountered is -95 dBm. Thus combining Figure 5 and Table 3 we conclude that even if the distance of the femto cells from the macro BS is 900 meters and beyond, the macro threshold has to be above -95 dBm to achieve higher load distribution to the femto cells.

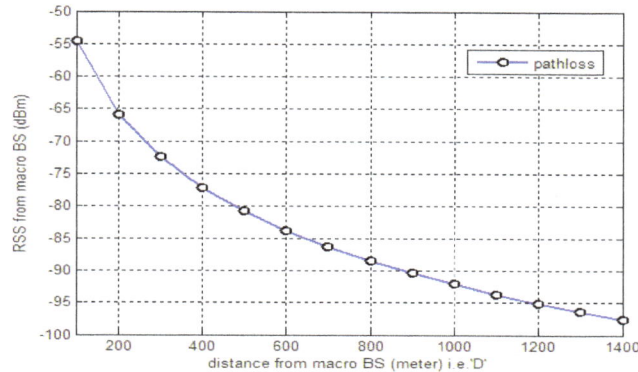

Figure 5. Pathloss from macro BS

Considering the above fact, we have observed the handoff probability for various kind of handoff decision discussed in this paper keeping $RSS_{m,th}$ at -70 dBm which is above the balanced threshold level for any 'R'. The result is shown in Figure 6.

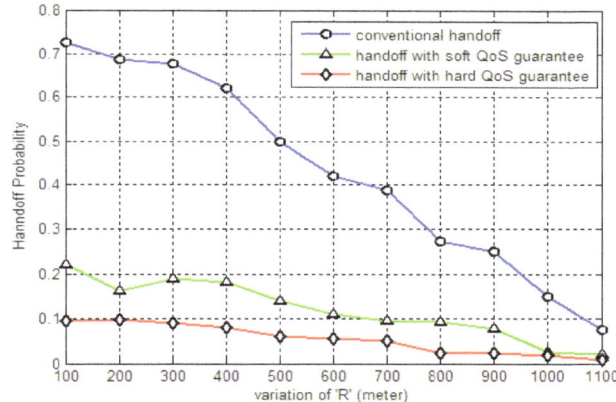

Figure 6. Handoff probability at $RSS_{m,th}$=-70dBm

Figure 6 reveals that the probability of handoff decreases to a considerable amount when unnecessary handoff is eliminated from the conventional handoff scheme. Again, the handoff probability in each case is also observed to fall gradually as 'R' increases. With increase in 'R', the outskirt region to be covered by the femto cells decreases. This in turn lowers the number of femto cells and thereby the handoff probability decreases. The handoff probability is observed to be much lower in case of hard QoS guarantee than the case of soft QoS guarantee. In a hierarchical cell scenario as shown in Figure 1, if a user moves with very high velocity or undergoes real-time call while moving from one end of the cell to the other end, reduction in the number of handoff is going to improve the call quality considerably. Thus soft QoS guarantee will ensure better call quality while hard QoS guarantee will provide much better call quality compared to conventional handoff scheme.

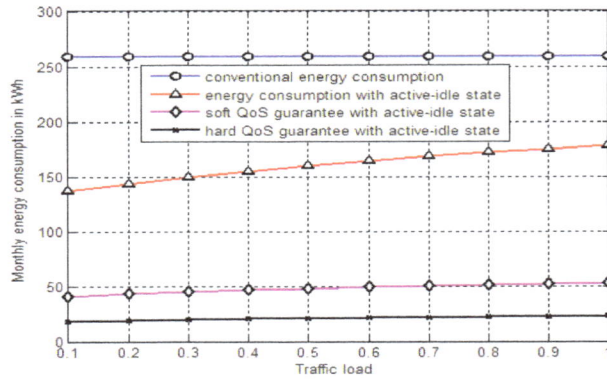

Figure 7. Monthly energy consumption for a single femto BS

Figure 7 shows the monthly energy consumption of the proposed handoff reduction technique considering the active-idle state of femto BSs with respect to the traffic load. As observed from the figure, the energy consumption remains unaffected with the variation of the traffic load for the conventional handoff strategy without consideration of the concept of active-idle state. However, the introduction of the active-idle state in the femto BSs conserves the monthly energy consumption to a huge extent. Energy is further conserved for the soft QoS and hard QoS guaranteed handoff reduction technique with the consideration of the active-idle state of femto BSs. Thus, the huge amount of energy is conserved by introduction of the active-idle mode of the femto BSs. The amount of energy saved is illustrated in Table 4.

TABLE 4. PERCENTAGE OF ENERGY SAVING

	Percentage of energy saving w.r.t conventional energy consumption				
Traffic Load	0.2	0.4	0.6	0.8	1.0
With active-idle state	47.81	43.70	40.33	37.58	35.32
Active-idle state with soft QOS	85.06	83.88	82.92	82.13	81.48
Active-idle state with hard QOS	92.76	92.19	91.72	91.34	91.03

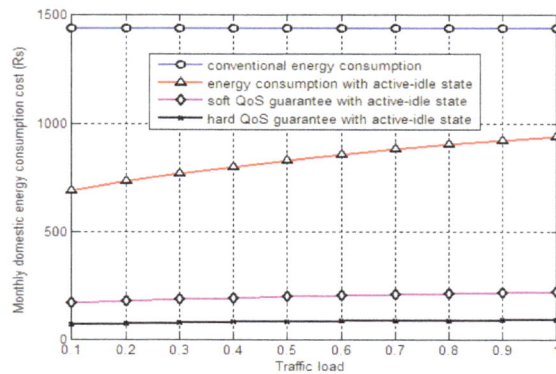

Figure 8. Monthly domestic energy consumption cost for a single femto BS

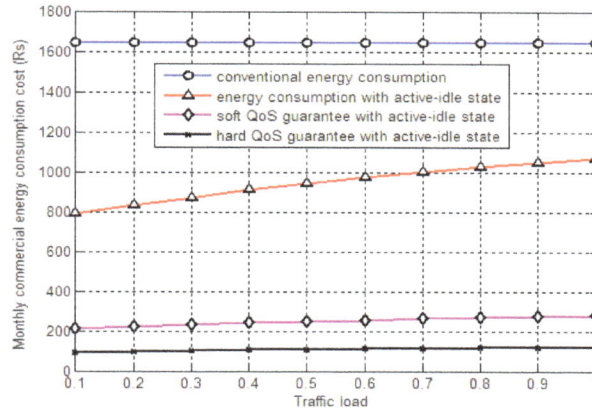

Figure 9. Monthly commercial energy consumption cost for a single femto BS

Based on the energy consumption tariffs of Calcutta Electricity Supply Corporation (CESC) of 2012 [17] the monthly energy consumption cost is calculated for domestic and commercial deployment of femto BS. Figure 8 and 9 reflects the domestic and commercial cost respectively for various handoff reduction techniques with active-idle state of femto BS. The profit gained for both the cases are summarized in Table 5 and 6.

TABLE 5. DOMESTIC PROFIT GAINED

	Percentage of domestic profit gained w.r.t conventional energy consumption				
Traffic Load	0.2	0.4	0.6	0.8	1.0
With active-idle state	52.62	48.10	44.39	41.36	38.88
Active-idle state with soft QOS	88.99	88.06	87.30	86.67	86.16
Active-idle state with hard QOS	94.88	94.47	94.14	93.87	93.65

TABLE 6. COMMERCIAL PROFIT GAINED

	Percentage of commercial profit gained w.r.t conventional energy consumption				
Traffic Load	0.2	0.4	0.6	0.8	1.0
With active-idle state	52.65	48.43	44.83	41.77	39.26
Active-idle state with soft QOS	87.70	86.73	85.94	85.29	84.76
Active-idle state with hard QOS	94.04	93.57	93.19	92.87	92.61

5. CONCLUSIONS

In this paper, we have proposed an energy efficient handoff decision algorithm for reducing unnecessary handoff in hierarchical macro/femto networks while balancing the load of macro and the femto BSs at minimal energy consumption. The performance of the proposed algorithm is also analyzed using CTMC model. Balanced threshold level of RSS from macro BS have been evaluated with respect to the distant location of the femto cells. Macro threshold level ($RSS_{m, th}$) set above the balanced threshold level results in higher load distribution of the mobile users to the femto cells. Hard QoS and soft QoS guarantee – the two decision of handoff reduction technique proposed in this paper shows how unnecessary handoff has been reduced while balancing the load of the macro and femto BSs. Soft QoS guarantee only will reduce the overhead of the network. On the other hand, hard QoS guarantee will reduce the network overhead as well as increase user satisfaction. So a service provider can choose any one of the QoS guarantee level depending upon their requirement. In addition, introduction of the active-idle state of the femto BSs shows significant reduction in the monthly energy consumption which is reflected in the domestic and commercial energy cost.

Since simulations are very time-consuming for broad scaled networks, the method of determining the balanced threshold level has been assessed for small scaled networks. However, this method can also be applied for broad scaled networks conceptually and the corresponding balanced threshold level can be calculated accordingly.

ACKNOWLEDGEMENTS

The authors deeply acknowledge the support from DST, Govt. of India for this work in the form of FIST 2007 Project on "Broadband Wireless Communications" in the Department of ETCE, Jadavpur University.

REFERENCES

[1] IEEE 802.16 Standard-(2001) "Local and Metropoliton Area Networks-Part 16", IEEE Draft P802.16/D3.

[2] Shu-ping, Y., Talwar, S., Seong-choon, L. & Heechang, K., (2008) "WiMAX Femtocells: A Perspective on Network Architecture, Capacity, and Coverage", IEEE Communications Magazine, Vol.46, No. 10, pp. 58-65.

[3] Kim, R. Y., Kwak, J. S. & Etemad, K., (2009) "WiMAX femtocell: requirements, challenges, and solutions", IEEE Communication Magazine, Vol.47, No. 9, pp.84-91.

[4] Zeng, H., Zhu, C. & Chen, W., (2008) "System performance of self-organizing network algorithm in WiMAX femtocells", Proceedings of the 4th ACM International Conference Proceeding Series. ICST, Brussels, Belgium, pp. 1-9.

[5] Lopez-Perez, D., Valcarce, A., De La Roche, G., Liu, E. & Zhang, J., (2008) "Access Methods to WiMAX Femtocells: A downlink system-level case study", 11th IEEE International Conference on Communication Systems, pp. 1657–1662.

[6] Claussen, H., (2007) "Performance of macro- and co-channel femtocells in a hierarchical cell structure", IEEE 18th International Symposium on PIMRC 2007, Athens, Greece, pp. 1–5.

[7] Halgamuge, M., et al., (2005) "Signal-based evaluation of handoff algorithms", IEEE Communication Letters, Vol. 9, No. 9, pp. 790–792.

[8] Hsin-Piao, L., Rong-Terng, J. & Ding-Bing, L., (2005) "Validation of an improved location-based handover algorithm using GSM measurement data" IEEE Transactions on Mobile Computing, Vol. 4, No. 5, pp. 530-536.

[9] Denko, M. K., (2006) "A mobility management scheme for hybrid wired and wireless networks", Proceedings of the 20th International Conference on Advanced Information Networking and Applications, Vol. 02, pp.366-372.

[10] Moon, J. & Cho, D. (2009) "Efficient handoff algorithm for inbound mobility in hierarchical macro/femto cell networks", IEEE Communications Letters, Vol.13, No.10, pp.755-757.

[11] Ashraf I., Ho L.T.W. & Claussen H., (2010) "Improving Energy Efficiency of Femtocell Base Stations via User Activity Detection" Proc. Of Wireless Communication and Networking Conference (WCNC), pp 1-5.

[12] Oh D.C., Lee H.C., Lee Y.H., (2010) "Cognitive Radio Based Femtocell Resource Allocation", International Conference on Information and Communication Technology Convergence (ICTC), pp. 274 – 279.

[13] Ross, S. M., (2001) "Probability Models for Computer Science", Elseveir, June.

[14] Baugh, C.R. & Huang, J, (2001) "Traffic Model for 802.16 TG3 Mac/PHY Simulations". IEEE 802.16 working group document.

[15] Ross, S. M., (1996) "Stochastic processes", John Wiley & Sons.

[16] Chandrasekhar, V., Andrews, J. & Gatherer, A. (2008) "Femtocell Networks: A Survey", IEEE Communication Magazine, Vol. 46, No. 9, pp. 59-67.

[17] http://www.cescltd.com/cesc/web/customer/regulator/document/CESC%20Tariff%20March%2012. pdf

9

Novel Cell Selection Procedure for LTE HetNets based on Mathematical Modelling of Proportional Fair Scheduling

Mohamed A. AboulHassan[1], Essam A. Sourour[2], Shawki Shaaban[2]

[1]Pharos University, Faculty of Engineering, Electrical Eng. Dept., Alexandria, Egypt
[2]Alexandria University, Department of Electrical Eng., Alexandria, 21544 ,Egypt

Abstract

Femtocells have been considered one of the most important technologies in LTE networks to solve indoor coverage problem, however the randomness deployment of femtocells, leads to great challenge for selecting optimum serving cell. In this work, a new cell selection algorithm is proposed that enables new user to select best serving cell whereas several factors are put into consideration other than highest instantaneous SNR or maximum RSRP such as cell load .A new prediction algorithm is designed to predict the performance of (PF) scheduling algorithm to calculate expected number of RBs to be scheduled to new user, then reduction in achievable data rate due to both received SNR and instant cell load is estimated. The numerical results show that the new proposed cell selection algorithm achieves higher average cell throughput than conventional cell selection methods and achieves less cell load variance between different adjacent cells.

Keywords

LTE; Proportional Fair scheduling algorithm, cell selection

1. Introduction

LTE (Long Term Evolution) [1] is a standard for wireless communication of high-speed data for mobile phones; it aims to achieve higher data rates meeting the rapid growth in demanded mobile applications. As, the percentage of traffic indoors has been increasing over time and is expected to increase further, good indoor coverage has been considered nowadays as one of the main design aspects to achieve target peak data rate, therefore low power nodes such as femtocells has been introduced as an effective solution. However, the existence of different network elements having distinct maximum transmit power that leads to the necessity of more complicated algorithm to achieve target peak data rate, Moreover, due to the shared technology of downlink shared channel in LTE networks, cell load participates in the parameters of the user's achievable data rate. Various Load balancing algorithms [2-4] have been proposed to maintain more balanced load for better average cell throughput. Other researches focus on improving the Resource Allocation (RA) procedure. Cell selection is considered one of important key points whereas optimum serving cell selection algorithms saves necessary downlink RA procedures, and load balancing operations. Conventional cell selection methods depending on highest instantaneous signal-to-noise ratio (SNR) or maximum reference signal received power (RSRP) have been studied to improve cell selection procedure , however such conventional schemes become inefficient due to the dependence of user's achievable data rate on cell load because of the shared resources technology implemented in the main LTE traffic channel (DPSCH), and the difference in transmitted power levels of LTE Heterogeneous Networks (HetNets) elements such as macrocells and femtocells. This may lead to degradation in user's average data throughput and consequently needs more dynamic cell selection algorithms. Several researches such as in [5-8] focus on improving conventional systems depending on maximizing instantaneous SNR or reference signal received power (RSRP) whereas in [9], the cell selection algorithms in heterogeneous networks are studied by considering the effect of different maximum transmitted power for different types of base stations. In [10] the effect of implementing Range expansion (RE) [11] is addressed. In order to get higher user's average throughput, cell selection algorithms have considered several factors other than received signal power level, as in [12] and [13], a long term analysis is investigated by considering the relation between cell load and instantaneous received signal level which is used by new user to select the best serving cell, such long term analysis is modelled as an optimization problem which needs a certain level of complexity that might not meet low latency targets.

In this paper, a novel cell selection algorithm is proposed by deriving a mathematical modelling of Proportional Fair (PF) scheduling algorithm [14] which is considered as one of the main scheduling algorithms used in LTE, the proposed algorithm can be applied for both short term and long term analysis, and doesn't need a complex operations as in algorithms based on optimization problems. First we design an algorithm to predict the behaviour of PF scheduling algorithm in future without a necessity to run the PF algorithm. The new prediction algorithm is then used by the new proposed cell selection algorithm to enable new coming user to select best serving cell with minimum expected reduction in achievable user's average data rate due to scheduling within existing users. We also design a standalone new proposed algorithm without a need for an extra overhead due to information exchange between different surrounding cells.

The rest paper is organized as follows; in section II new algorithm for modelling PF scheduling algorithm is proposed. Section III describes the new cell selection algorithm depending on PF Prediction algorithm presented in section II. Simulation results are shown in section IV. Finally section V yields concluding remarks.

2. NEW PROPOSED PREDICTION ALGORITHM

2.1. PF scheduling Algorithm

   LTE mobile networks rely on scheduling available resources among existing users to achieve an acceptable compromise between fairness and individual user's throughput, therefore the base station needs to run appropriate algorithms to take decisions periodically for assignment of available resources. The minimum resource unit can be assigned to a single user is called Resource Block (RB). One of the most commonly used scheduling algorithms is Proportional fair (PF) algorithm. Previous research as in [15] and [16] has shown that PF scheduling algorithm achieves good balance between user's fairness and average cell throughput, therefore PF algorithm is considered as a standard scheduling algorithm in LTE networks. The work in [17] and [18] has focused on finding a closed form for the user's throughput and system throughput, nevertheless we still need to predict the pattern of served users implementing PF algorithm

The PF depends on selecting the user algorithm having the metric M defined as

$$M = arg\ max_i\ \left(\frac{r_i}{R_i}\right) \tag{1}$$

And

$$R_i(t+1) = \begin{cases} R_i(t) * \left(1 - \frac{1}{t_c}\right) & user\ i\ is\ scheduled \\ R_i(t) * \left(1 - \frac{1}{t_c}\right) + r_i(t) * \frac{1}{t_c} & user\ i\ is\ not\ scheduled \end{cases} \qquad (2)$$

Where $Ri(t)$ is the average moving throughput of user i over a time window (t_c) of an appropriate size, $R_i(t+1)$ is updated value after every assigned RB. $r_i(t)$ is the achievable data rate of user i and can be calculated as follows [19]:

$$r_i(t) = \frac{nbits_{i,j}(t)}{symbol} * \frac{nsymbols}{slot} * \frac{nslots}{TTI} * \frac{nsc}{RB} \qquad (3)$$

Values of $r_i(t)$ can be obtained from mapping the values of instantaneous Signal-to-Noise Ratio (SNR) into values of achievable Data rate as in table 1 [20]

SNR is given by:

$$SNR = \frac{P_t}{Rbno} \times \frac{10^{-\frac{SHL}{10}}}{(I+N)} \qquad (4)$$

Where P_t is the maximum transmitted power value , $Rbno$ is number of RBs per TTI , PL is the path loss value ,SHL is Shadowing fading , N is the white noise power and I is the inter-cell interference from other cells. In the following subsection, we will propose the new prediction algorithm that estimates the pattern of scheduled users i.e. the ID of scheduled users within each RB without the need to run the PF algorithm and updating vales of average moving throughput R for each user after each RB

Table 1. values of SNR and their corresponding achievable data rate

Minimum Instantaneous Downlink SNR Value (dB)	MCS	Data Rate (kbps)
1.7	QPSK(1/2)	168
3.7	QPSK(2/3)	224
4.5	QPSK(3/4)	252
7.2	16QAM(1/2)	336
9.5	16QAM(2/3)	448
10.7	16QAM(3/4)	504
14.8	64QAM(2/3)	672
16.1	64QAM(3/4)	756

2.2. System model

Assume a base station (BS) consisting of U active users, each user has initial average moving throughput R_i and achievable data rate r_i where $i=\{1,2..U\}$,the ID of new user is $U+1$. The user's scheduling rate S_i of user i is defined as

$$S_i = \frac{r_i}{R_i} \qquad (5)$$

It can be concluded that under the assumption of flat fading channel and low mobility environment, the value of instantaneous SNR in (4) depends mainly on path loss value which remains almost constant for a certain interval because the log normal shadowing has zero mean, so without loss of generality SNR values assumed to be constant through a certain prediction interval, moreover other assumptions as in [18] considered an average values for measured SNR for long term analysis.

The prediction interval can be divided into two periods:

i) *Period(1)* :considered as transient period where only some of active users are scheduled while other users are not scheduled.

ii) *Period(2)* : considered as steady state period where all active users are scheduled to be served periodically i.e. performance of PF algorithm tends to be the same as Round Robin (RR) scheme

Due to the shared downlink technology of downlink shared LTE main traffic channel, the achievable Downlink Data rate depends mainly on cell load and scheduling criteria, thus we need to estimate the number of RBs assigned to the new user to find the effective achievable Data rate i.e. Practical data rate of new user with certain achievable data rate after being scheduled within a certain cell. To calculate the number of RBs expected to be assigned to new user, we have to calculate the number of RBs assigned to a new user within transient period and number of RBs assigned in steady state period as follows

$$n_{u+1}^{RB} = n_{P1}^{RB} + \frac{(n_{total}^{RB} - \sum_{l=1}^{k} m_k)}{U+1} \tag{6}$$

Where $n_{p1}{}^{RB}$ is number of RBs assigned to new user in transient period, $\sum_{l=1}^{k} m_k$ is total number of RBs to be scheduled in transient period and $n_{total}{}^{RB}$ is the total number of RBs to be scheduled during prediction interval and equals to

$$n_{total}^{RB} = n_{TTI}^{RB} * N_{sf} * N_{fr} \tag{7}$$

Where n_{TTI}^{RB} is number of RBs per TTI, N_{sf} is number of subframes/frame and N_{fr} is number of frames within prediction interval. The second term in equation (6) represents number of RBs assigned in the steady state period, assuming new user to be served in Round Robin (RR) pattern.

Let r_{U+1} is the new coming user's achievable data rate and R_{U+1} is the new user's average moving throughput ,assumed to be minimum as new user has no history. To get the number of RBs in transient period, we need to predict the new serving pattern of all users i.e. ID of user served in each RB during prediction interval after new user is scheduled. The transient period can be divided into number of phases where each phase k contains m_k users, sorted in descending order according to scheduling ratio S where

$$m_k = \{m_k \leq U \mid S_i^d > S_j^g \text{ for all } i \text{ and } j \leq m_k \text{, } d \text{ and } g \text{ are the order of } S \text{ in phase } k,$$

$$\text{and } d < g \} \tag{8}$$

The end user of each phase k is defined as: the user having minimum S and has been served in previous phase k-1.In classical PF algorithm, the value of moving average throughput R for each user is updated after each scheduling decision is taken i.e. after each RB, which will consumes a lot of time and calculations, such massive calculations are avoided in the new prediction algorithm by two main methods:

1- The average moving throughput value for each user is updated after the end of each phase instead of each RB by using equation (9)

$$R_u(l) = R_u(0)f^N + r_u(1-f)(\textstyle\sum_{k=1}^{N} I_{uk}f^{N-k}) \tag{9}$$

Where

l : current phase ID

N: total number of RBs shceduled in previous phases

f: constant factor equals $(1 - \frac{1}{t_c})$, where $f < 1$ i.e. $t_c > 1$

I_{uk}: elements of serving pattern matrix (SPM) having binary value i.e.

$$I_{uk} = \begin{cases} 1 & \text{if user } u \text{ served in RB } k \\ 0 & \text{otherwise} \end{cases}$$

SPM is a m_{SPM} x n_{SPM} matrix whose entries are binary value indicating whether user is scheduled or not , $m_{SPM} = U+1$ and $n_{SPM} = n_{total}^{RB}$.

$$SPM_{m_{SPM}n_{SPM}} = [I_{uk}] \tag{10}$$

where each column in SPM has only one value =[1] and the remaining values =[0] because only one user is scheduled in each RB.

2- The user pattern is predicted only in the transient interval while in steady state period, the number of RBs expected to be assigned to new user is calculated by dividing number of RBs in steady state period by number of users due to the fact that the performance of PF algorithm is similar to Round Robin scheme as it will be proven later..

2.3 Mathematical modelling of prediction algorithm

Assume active user has initial value $R_i(0)$,and new user has initial value $R_{u+1}(0)$ assumed to be minimum where $R_{u+1}(0) \le R_i(0)$ for all $i=1...U$ The following theorems are stated to verify the procedure of the proposed prediction algorithm

2.3.1. Theorem 1:

This theorem states that Phase 1 always contains one user which is new coming user

Proof

New coming user has no history regardless its achievable data rate ,therefore R_{U+1} has minimum value that leads to maximization of S_{U+1},therefore

$$S_{U+1}(1) \ge S_j(1) \qquad \forall j=1.....U \tag{11}$$

where $S_{U+1}(1)$ is the scheduling rate of user $u+1$ in phase 1.

2.3.2. Theorem 2:

This theorem states that new user $U+1$ is always the last member in every scheduling set of every phase

Proof

By definition, the members of each phase are determined by users having updated values of $S_j > S_{u+1}$ and as stated in theorem 3 the order of users in each phase is not changed, therefore new user is the last user in each phase

2.3.3. Theorem 3 :

This theorem states that for any two users i and j having S_i and S_j respectively where $S_i(k) > S_j(k)$ in phase k, then $S_i(k+1) > S_j(k+1)$ for all values of i,j and k

Proof

N.B. without loss of generality we will apply the proof on transition from phase 2 to phase 3(phase 1 contains only one user as per theorem 1) to trace the behaviour of users then we will show that it can be generalized on all phases

Consider phase 2 has m_2 users, *user$_j$* is the first member in phase two i.e. highest value $S_j^{(m_2)}(2)$ after end of phase two, *user U+1* is the last member in phase two having $S_{U+1}^{(m_2)}(2)$, at end of phase 2 the values of S are

$$S_j^{(m_2)}(2)= \frac{S_j}{f^{m_2+1}+S_j(1-f)(f^{m_2-1})} \tag{12}$$

$$S_{U+1}^{(m_2)}(2)= \frac{S_{U+1}}{f^{m_2+1}+S_{U+1}(1-f)(1+f^{m_2})} \tag{13}$$

and

$$S_j^{(m_2)}(2)> S_{U+1}^{(m_2)}(2) \tag{14}$$

then

$$\frac{S_j}{f^{m_2+1} + S_j(1-f)(f^{m_2-1})} > \frac{S_{U+1}}{f^{m_2+1} + S_{U+1}(1-f)(1+f^{m_2})} \tag{15}$$

We assume that S_j is the initial value of scheduling rate of users, the superscript indicates the number of RBs in each phase after which S is updated, and we omit the phase ID for simplicity.

To prove theorem 3, we have to prove that the inequality $S_j^{(1)}(3) < S_{U+1}^{(1)}(3)$ is impossible, where the inequality indicates values of S after assigning first RB in phase 3 to *user$_j$*

$$\frac{S_j}{f^{m_2+2} + S_j(1-f)(1+f^{m_2})} < \frac{S_{U+1}}{f^{m_2+2} + S_{U+1}(1-f)(f+f^{m_2+1})} \tag{16}$$

$$\therefore S_{U+1}S_j(1-f)^2(1+f^{m_2}) < \left(S_{U+1} - S_j\right)f^{m_2+2} \tag{17}$$

Which is impossible because from (11) RHS < 0 and LHS > 0 by definition

We can notice that for any two phases k and $k+1$, there will be value m_{k+1} in the power term of LHS, using binomial expansion, this value doesn't lead to change sign of LHS, Therefore we can generalize rule three for any phase k

2.3.4. Theorem 4:

This theorem is considered as more generalization for theorem 3, it states that the order of served m_k users in phase k is preserved during all incoming phases

Proof

At end of phase 1

$$S_j^{(1)}(1) > S_{U+1}^{(1)}(1) \tag{18}$$

where the probability of one user at least will have updated S value greater that the update S value of new user is high due do the range of values of r.

then

$$\frac{S_j}{f} > \frac{S_{U+1}}{f + S_{U+1}(1-f)} \tag{19}$$
$$\therefore (S_{U+1} - S_j)f < S_j S_{U+1}(1-f) \tag{20}$$

Then after *user j* is served in phase two and *user$_{U+1}$* is in last RB in phase two. To prove theorem 4, we will show that the inequality in (21) is impossible,

At the end of phase 2 if the order is changed

$$\therefore S_j^{(m_2)}(2) < S_{U+1}^{(m_2)}(2) \tag{21}$$
$$\therefore \frac{S_j}{f^{m_2+1} + S_j(1-f)f^{m_2-1}} < \frac{S_{U+1}}{f^{m_2+1} + S_{U+1}(1-f)(1+f^{m_2})} \tag{22}$$
$$\therefore (S_{U+1} - S_j)f > S_j S_{U+1}(1-f)\frac{(1+f^m - f^{m-1})}{f^m} \tag{23}$$

it is obvious that using binomial expansion, the $\frac{(1+f^m - f^{m-1})}{f^m} > 1$ for all values of m, therefore from (23), equation (21) is impossible. Therefore the order is preserved regardless the length of the phase i.e. for any values m_k, the order is preserved

2.3.5. Theorem 5:

This theorem states that, based on theorems *1, 2, 3* and *4*, for phase k having m_k members equals $U+1$, all users are served in periodic pattern i.e. Round Robin (RR) in steady state period with period length $U+1$ and *user $U+1$* is the last member of the period. Under assumption of flat fading channel and low mobility environment, users are then served in a Round Robin (RR) pattern until the end of the prediction interval which can be considered steady state period so no need to run any simulation to calculate the number of RBs assigned to new user for the rest of the prediction interval. This result complies with the results in [18] which proved by simulation that the values of R tend to be constant after certain time.

In LTE networks, users are served through Downlink Physical Shared CHannel (DPSCH) where RBs are assigned to users according to scheduling algorithms. Therefore the average user rate depends not only on instantaneous SNR but also on number of RBs assigned per frame. In the new cell selection proposed algorithm, new user calculates the theoretical achievable data rate based on instantaneous SNR and receives Reduction Factor (RF) then calculates the maximum effective achievable Data Rate.Formula in (3) stated that according to instantaneous SNR, new user can attain an achievable data rate from table 1 assuming that user is scheduled one per subframe. After calculating number of RBs expected to be assigned to new user, Reduction Factor (RF) due to cell load can be calculated as shown

$$\text{RF} = \frac{n_{u+1}^{RB}}{N_{sf} * N_{fr}} \qquad (24)$$

Where n_{u+1}^{RB} is obtained from (6) and N_{sf} is number of subframes/frame and N_{fr} is number of frames during prediction interval .It can be noticed that values of RF depends on criteria of mapping SNR values into data rates as in Table 1 whereas values in Table 1 can be achieved if the user is scheduled once every subframe. Figure 1 summarizes the new proposed prediction algorithm.

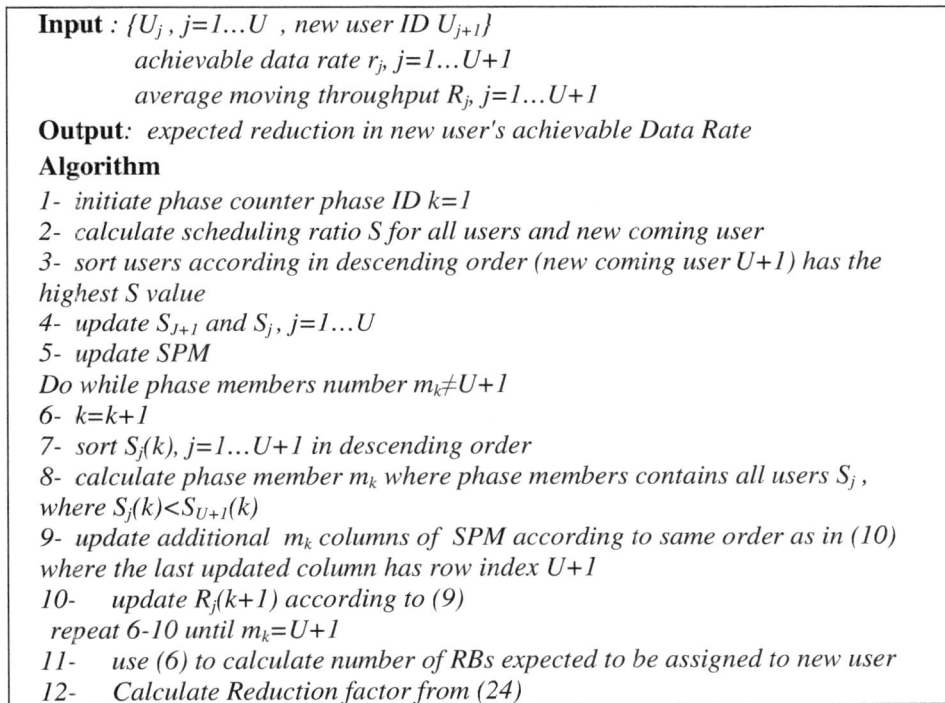

Input : $\{U_j, j=1...U$, new user ID $U_{j+1}\}$
        achievable data rate $r_j, j=1...U+1$
        average moving throughput $R_j, j=1...U+1$

Output: expected reduction in new user's achievable Data Rate

Algorithm

1- initiate phase counter phase ID $k=1$
2- calculate scheduling ratio S for all users and new coming user
3- sort users according in descending order (new coming user $U+1$) has the highest S value
4- update S_{J+1} and $S_j, j=1...U$
5- update SPM
Do while phase members number $m_k \neq U+1$
6- $k=k+1$
7- sort $S_j(k), j=1...U+1$ in descending order
8- calculate phase member m_k where phase members contains all users S_j, where $S_j(k) < S_{U+1}(k)$
9- update additional m_k columns of SPM according to same order as in (10) where the last updated column has row index $U+1$
10- update $R_j(k+1)$ according to (9)
  repeat 6-10 until $m_k=U+1$
11- use (6) to calculate number of RBs expected to be assigned to new user
12- Calculate Reduction factor from (24)

Figure.1: Summary of modelling of PF scheduling algorithm

Hence the effective Data rate of the new user for each of the surrounding cells is given by

$$effective\ DR = RF * r_{u+1} \qquad (25)$$

To verify the accuracy of the prediction algorithm, we perform a simulation to predict the effective data rate of new user using classical prediction algorithm implementing PF criteria after each RB for the same prediction interval as in Fig. 2, we assume here a single cell simulation consisting of several number of users, the value of effective data rate for new user is calculated for all possible values of data rates, the users are assumed to be disturbed uniformly within the cell coverage area. It can be shown that for constant SNR values during the prediction interval, proposed prediction algorithm outputs the same values of effective data rate for new user as a classical prediction algorithm that needs a massive calculation especially with long term prediction.

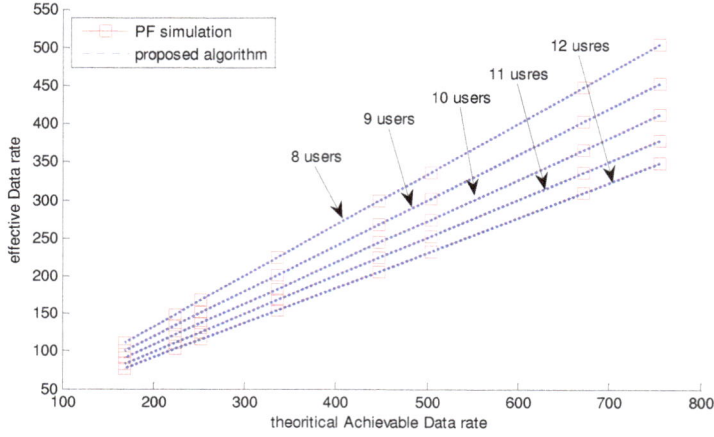

Figure 2. Comparison between classical prediction algorithm and proposed algorithm

2.4. Throughput analysis

When studying throughput analysis, we can notice that the total throughput during the simulation time is affected by the new user achievable data rate where the total cell throughput can be increased or decreased as shown in Fig. 3. The total cell throughput for single cell consisting of 8 uniformly distributed users is calculated before and after the new user is being scheduled. It can be shown that the new user can lead to increase or decrease the total cell throughput according to the new users achievable data rate, therefore a new constrain is added to our cell selection criteria to maintain both individual user improved data rate and total cell throughput. The total cell throughput is directly proportional to achievable data rate of the user, so it is preferred that new user's achievable data rate achieves total throughput greater that the total throughput before scheduling new user, the predicted total cell throughput is given by

$$\text{average cell throughput} = \sum_i^{u+1} (n_i^{RB} * r_i)/(N_{fr} * N_{sf}) \tag{26}$$

The achievable data rate is calculated with assumption that user is served once per subframe i.e. once every *1 ms*

Therefore

$$user\ throughput_i = \frac{n_i^{RB} * r_i}{20 n_f} \tag{27}$$

$\therefore$ average cell throughput before sceduling new user $= \sum_{i=1}^{u} \frac{n_i^{RB} * r_i}{20 n_f}$ (28)

the following inequality has to be fulfilled to increase total throughput

average cell througput after scheduling new user
> average cell througput before scheduling new user

then

$$\sum_{i=1}^{u} \frac{n_i^{RB} * r_i}{20 n_f} < \sum_{i=1}^{u+1} \frac{m_i^{RB} * r_i}{20 n_f} \tag{29}$$

where

n^{RB}: number of RBs assigned to user i before new user is scheduled
m^{RB}_{i}: number of RBs assigned to user i after new user is scheduled
n_f :number of frames
number of RBs/subframe=2
number of subframes / frame=10

For long simulation time i.e. large number of RBs, we need to calculate a lower bound, depending on rule 5 we can neglect the number RBs assigned to user within transient period, number of RBs assigned to user can be approximated to RR performance and equals to

$$\frac{total\ number\ of\ RBs}{number\ of\ users} = \frac{12*10*n_f}{u} \tag{30}$$
$$where\ n_i^{RB} \cong \frac{120*n_f}{u}\ and\ m_i^{RB} \cong \frac{120*n_f}{u+1}$$

Then

$$\sum_{i=1}^{u} \frac{120n_f*r_i}{u*n_f} < \sum_{i=1}^{u+1} \frac{120n_f*r_i}{(u+1)n_f} \tag{31}$$

$$r_{u+1} > \frac{\sum_{i=1}^{u} r_i}{u} \tag{32}$$

Therefore the achievable data rate of the new user must be greater that $\frac{\sum_{i=1}^{u} r_i}{u}$ so that the new user scheduling yields to an expected increase in total cell throughput

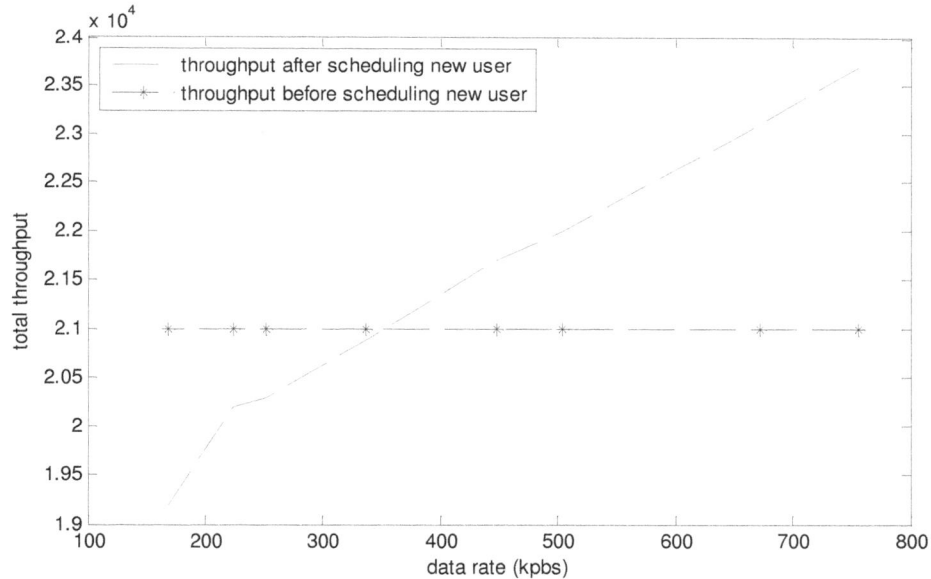

Figure3. Effect of new user's achievable data rate on cell average throughput

3. PROPOSED CELL SELECTION ALGORITHM

In this section, we propose two schemes for our new proposed cell selection algorithm, both schemes use prediction algorithm as a main algorithm for calculating the expected number of RBs, however first scheme can need a to be implemented partially in a central node so it can be considered as a hybrid scheme while the second scheme is considered as a standalone algorithm.

3.1. Proposed Hybrid Cell Selection algorithm (HCSA)

The target of the proposed cell selection algorithm is to calculate the reduction in achievable Data rate due to cell load, therefore we need to determine the values of SNR measured by new user with respect to all surrounding cells then simulate the behaviour of user within each surrounding cell to calculate the maximum effective data rate. In the hybrid proposed cell selection algorithm, the procedure is divided between the new user and a central node where the new user measures the value of instantaneous SNR w.r.t all surrounding cells. Through one of surrounding cells, the central node is fed back by the information of each cell load i.e. number of users and their SNR values then the central node runs the proposed prediction algorithm to estimate the effective data rate of the new user for each of the surrounding cells by using equation (25).

The central node sends the ID of the cell achieving maximum effective Data rate to the new user preserving the cell throughput constrain in (32). Therefore the new user transmits instantaneous SNR w.r.t all surrounding cells to a central node via one of surrounding cells, central node calculates the expected reduction in achievable data rate of the new user for each cell then reports the reduction to the new user. The new user selects the cell providing the user with maximum effective achievable data as follows

$$Cell\ selected\ (c) = arg\ max_{(c)}\{RF_{u+1}{}^{c}*\ r_{u+1}{}^{c}\} \qquad s.t.\ r_{u+1} > \frac{\sum_{i=1}^{u} r_i}{u} \qquad (33)$$

Proposed cell selection algorithm is summarized in Fig. 2

Algorithm

1- *New user measures received SNR from all surrounding cells*
2- *New user transmits received SNR values to a central node*
3- *Central node runs proposed prediction algorithm in section 2 to estimate the behavior of new user within all reported surrounding cells and calculates the expected reduction in theoretical achievable data rate between new user and each of surrounding cell as in (24)*
4- *The reduction factor values are sent to new user*
5- *New user calculates effective Data rate according to (25)*
6- *New user selects the cell achieving maximum effective Data rate fulfilling the criteria in (33)*

Figure 4. Summary of HCSA.

3.2. Proposed Standalone Cell Selection Algorithm (SCSA)

Hybrid cell selection algorithm needs some sort of coordination and signalling overhead between new user and serving cells and also an existence of a central node which will causes an overhead over the signalling links. Considering that users are served according to the maximization of scheduling ratio S and values of average moving throughput R is updated according to (2).

We can apply a constant scaling factor to all values R_i such that $R_i << r_i$, then the values of S after substituting (2) in (5) will be almost independent of values of r , so each cell can calculate an expected reduction factor regardless the achievable of new user by using an arbitrary value for achievable data rate of new user. Therefore signalling overhead can be omitted whereas each cell calculates its expected reduction factor for new user's achievable data rate then broadcasts it to the new user. New user then calculates effective Data Rate according to measured SNR and received reduction factor values without any necessity for processing in central node or data exchange, however the constrain in (31) cannot be fulfilled in standalone algorithm because it needs data exchange between adjacent cells. Fig.3 summarizes the standalone new proposed cell selection algorithm.

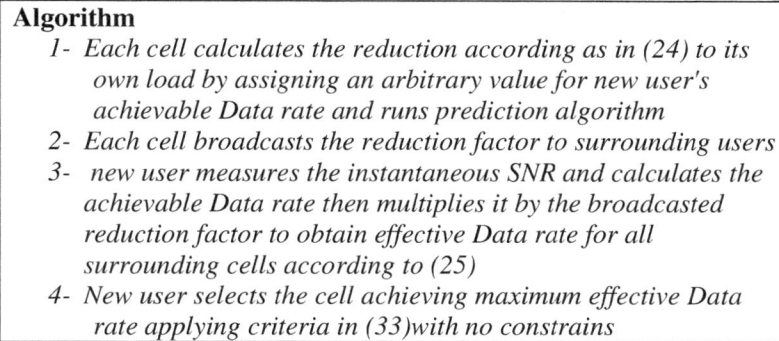

| **Algorithm** |
| 1- *Each cell calculates the reduction according as in (24) to its own load by assigning an arbitrary value for new user's achievable Data rate and runs prediction algorithm* |
| 2- *Each cell broadcasts the reduction factor to surrounding users* |
| 3- *new user measures the instantaneous SNR and calculates the achievable Data rate then multiplies it by the broadcasted reduction factor to obtain effective Data rate for all surrounding cells according to (25)* |
| 4- *New user selects the cell achieving maximum effective Data rate applying criteria in (33)with no constrains* |

Figure 5. Summary of SCSA.

3.3 Simple Prediction Cell Selection Algorithm(SPCSA)

We propose a simple cell selection algorithm similar to SCSA however the calculations of the reduction factor assume new user will be served in a RR pattern through the entire prediction interval, therefore the reduction factor is calculates as

$$RF_{SPCSA} = \frac{RBno}{U+1} \qquad (34)$$

Where *RBno* is thenumber of RBs per TTI, then effective data rate can be calculated using equation (25)

4. SIMULATION RESULTS

A multi-cell system level simulation using Matlab for a LTE network is performed in this section, as illustrated in Fig. 6, for a hexagonal cell layout with 7 cells and 3 sectors/cell, each are 120^0 apart. Two femtocells are assumed to be randomly located in each sector. Table 2 summarizes simulation parameters. Six cell selection schemes are investigated as shown in Table 3. We first simulate the behaviour of a single user attempting to perform a cell selection over an existing network, the new user is assumed to be located within the coverage area of one of the existing femtocells. Figure 7 plots CDF of the average throughput of the new user. We notice here that the new proposed schemes (scheme *4* and scheme *5*) outperform conventional schemes (scheme *1* and scheme *2*) by approximately 30% increase in average user's throughput , however scheme 3 and scheme 6 achieves almost the same performance as proposed schemes(scheme *4* and scheme *5*). When investigating the rate of femtocell selection, new proposed schemes could have a moderate and acceptable femtocell selection percentage than conventional schemes, where schemes 1 and 2 have a very low femtocell selection rate while scheme 3 have a very high rate which is considered as an advantage due to small coverage area of femtocells and overlaying coverage are between macrocells and femtocells as shown in Table 4. In Figure 8, we investigate the new user's average throughput where the user is assumed to be located in a macro cell coverage area, the proposed schemes (scheme *4* and scheme *5*) achieve higher Data rates than conventional schemes (schemes 2 and 3) by 14% and 6% respectively while the performance of scheme 1 has almost the same average throughput as proposed schemes, still scheme 6 has the same performance as proposed schemes (schemes 4 and 5) , we also notice here that scheme 3 that shows a good performance in case user is in femtocell coverage while its performance shows a degradation in case user is located in the coverage area of a macrocell .

Figure 9 examines an entire cluster applying addressed cell selection schemes, it can be shown that new proposed schemes achieves about 12% and 3 % improvement in average cell throughput than scheme 2 which is considered as the standard cell selection algorithm in LTE systems, and scheme 3 respectively. The performance of scheme 1 is almost similar to new proposed schemes (schemes *4* and *5*). The previous improvements remain almost unchanged when calculating the average cluster's macrocell throughput as in Fig 10. Despite the similarity of average throughput between proposed schemes and schemes 3 and 1 in case of femtocell selection and macrocell selection respectively, the new proposed schemes achieves better fairness among all cells whereas the variance between number of macrocell users in different cells is minimized in case of new proposed schemes as in Fig. 11 which reduces the necessity of Load Balancing algorithms between adjacent cells. Scheme 4 has the best various among all addressed schemes where it has less variance than scheme 5,6 and 3 by almost 8% and less than schemes 1 and 2 by almost 22% which indicates that proposed scheme 4 achieves the best fairness among adjacent cells as users are assumed to be distributed uniformly within the macrocell coverage area.

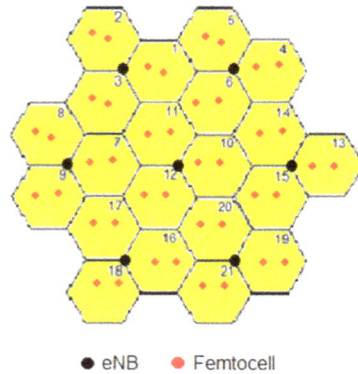

● eNB   ● Femtocell

Figure 6. Cell layout.

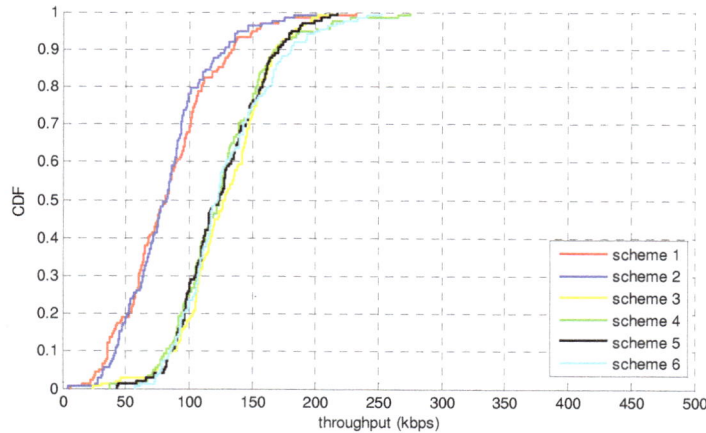

Figure 7. New user's average throughput (located in femtocell coverage area)

Table 2. Simulation parameters

Parameter	Setting
System bandwidth	1.2 MHz
Carrier frequency	2GHz
No. of RBs per half TTI(*0.5 ms*)	6
Cell layout	Hexagonal layout,3 sector per site,resue1
Number of sites	7 (=21 cell)with wrap-around
Macrocell transmit power	43 dBm
Intersite distance	350 m
Max number of users per Macrocell	50
Macrocell BS Antenna gain	14 dBi
Dependent path loss (PL) from Macrocell BS to Macrocell MS	$14.2+37.6\log_{10}$ (D in m)
Dependent path loss (PL) from Macrocell BS to femto MS	$14.2+37.6\log_{10}$ D+PeL(D in m) PeL: Penetration Loss
White noise power density	-174 dBm/Hz
Shadowing fading (SHL)	Log-normal distribution
SHL standard deviation	8 dB
User velocity	3 Km/s
Penetration Loss (PeL)	10 dB
Femtocell Radius	20m
No. of femtocells BS per macrocell	2
Min. distance between Macrocell and Femtocell	35 meters
Dependent path loss (PL) from Femtocell BS to femtocell MS	$38.46+ 20 \log_{10}$(D in m) + 0.7*d [dB] d [m]:distance inside house (assumed to be 2 m)
Dependent path loss (PL) from Femtocell BS to Macrocell MS	$Max(14.2+37.6\log_{10}$ (D), $38.46+ 20 \log_{10}$(D)+ 0.7*d+ PeL)
Path loss from femtocell BS to femtocell MS in another femtocell	$Max(14.2+37.6\log_{10}$ (D), $38.46+ 20 \log_{10}$(D)+ 0.7*d+2* PeL)
Shadowing fading standard deviation (indoor environment)	4 dB
Femtocell transmit power	10 dBm
Maximum no. of femtocell users	10

Table 3. Simulation scenarios

Scheme no	Scheme type
Scheme 1	Highest SINR
Scheme 2	Highest RSRP
Scheme 3	Minimum path loss
Scheme 4	Proposed HCSA
Scheme 5	Proposed SCSA
Scheme 6	Proposed SPCSA

Table 4. Percentage of femtocell selection

Scheme ID	Percentage of femtocell selection
Scheme 1	21.8%
Scheme 2	15%
Scheme 3	91.2%
Scheme 4	36.4%
Scheme 5	`38.6%
Scheme 6	48.9%

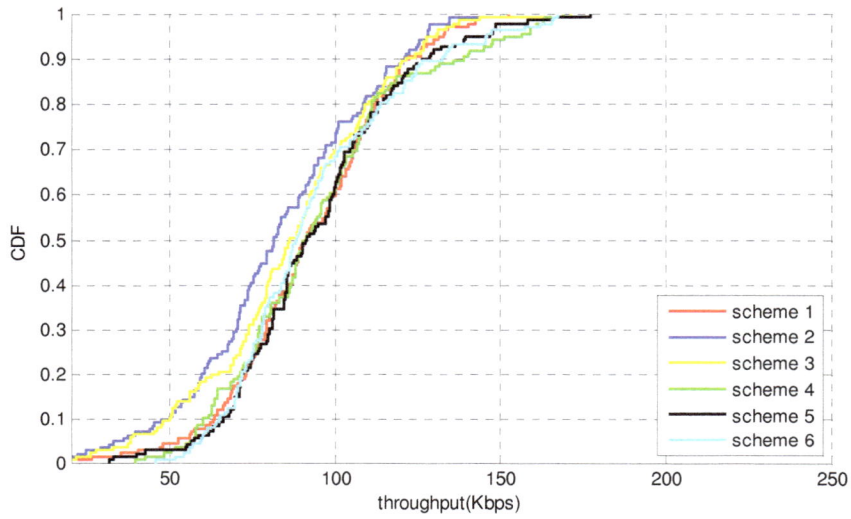

Figure 8. New user's average throughput (located in macrocell coverage area)

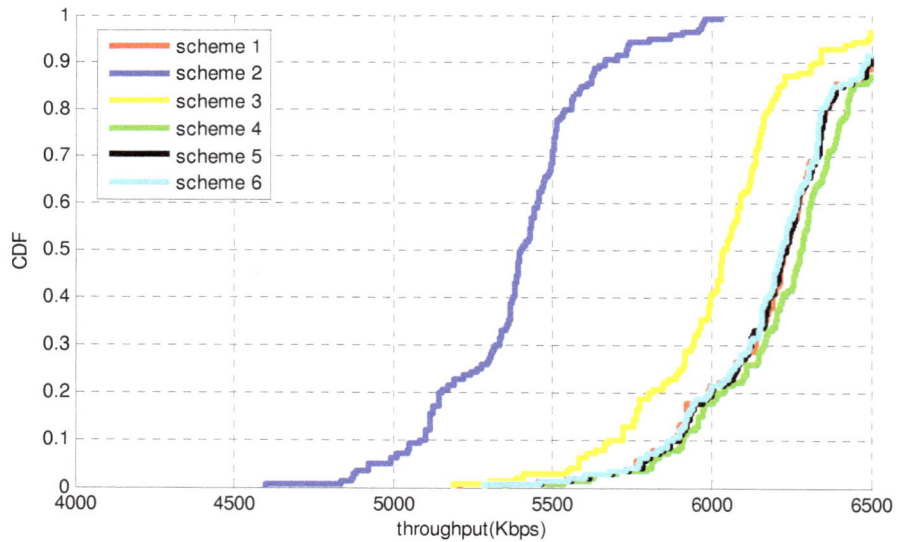

Figure 9. Average cell throughput of addressed cluster

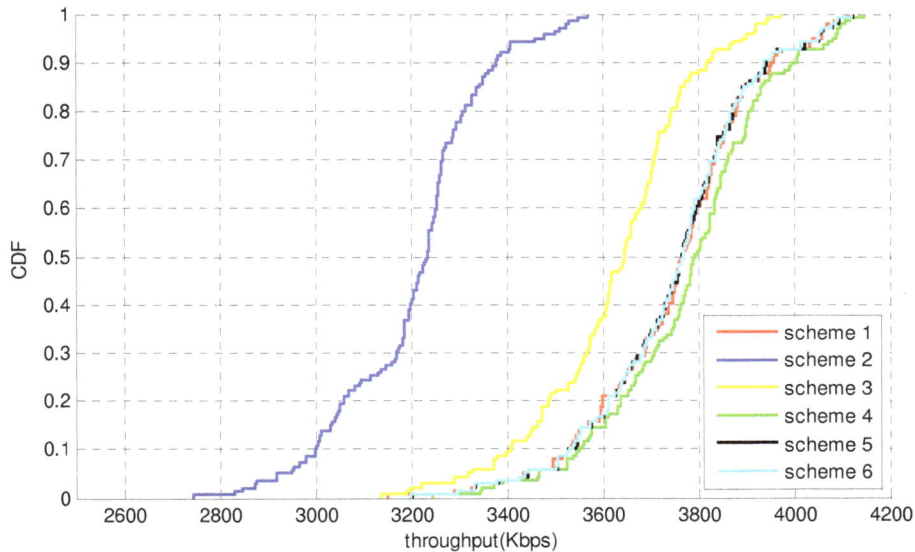

Figure 10. Average macrocell throughput of addressed cluster

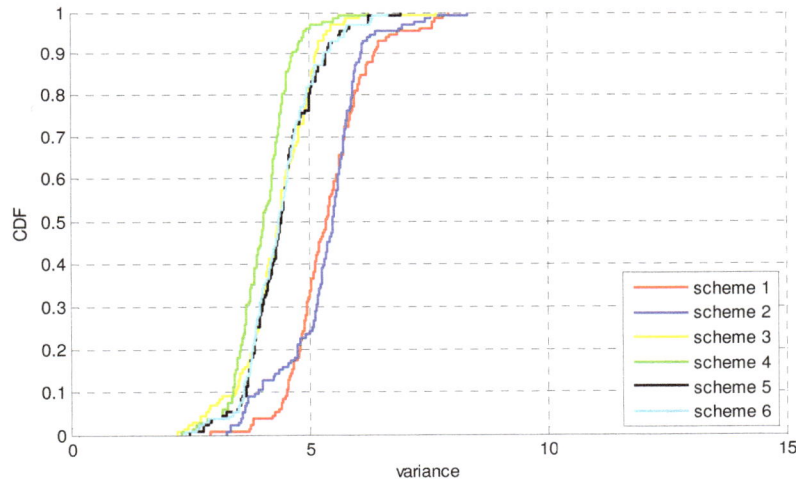

Figure 11. variance between numbers of users per macrocell

5. CONCLUSION

In this paper, we studied the cell selection operation in LTE Heterogeneous Networks, firstly we designed a prediction algorithm to estimate the expected number of RBs that will be assigned to a new user for a certain predetermined prediction interval without need to apply PF algorithm after each RB for the entire interval, then a new cell selection algorithm is proposed to calculate the expected reduction in new coming user's achievable data rate. 3 schemes of our new proposed cell selection algorithm have been addressed. The new proposed schemes can be implemented in both hybrid and standalone modes. Simulation results show that the new proposed schemes achieves better values for both user's average Data rate and average cell throughput than conventional scheme 2 which is considered as the standard LTE cell selection algorithm, also other conventional cell selection algorithms fail to achieve the best performance in both femtocell and macrocell environment where scheme 3 achieves almost a good performance as proposed schemes in case user is located in a femtocell coverage area while scheme 1 achieves a good performance as proposed schemes if users are located in macrocell coverage area, proposed schemes maintain their outperformed performance in both cases. In addition to improved average throughput, proposed scheme 4 has the best fairness among adjacent cells and thus reduces the need for load balancing algorithms that need a lot of time consuming complex algorithms and signalling overhead.

REFERENCES

[1] 3GPP, www.3gpp.org.

[2] H. Zhang ; X. Qiu ; L. Meng ; X. Zhang, (2010)"Design of Distributed and Autonomic Load Balancing for Self-Organization LTE ",in IEEE 72nd VTC 2010-Fall ,pp. 1 – 5.

[3] H. Wang ; L. Ding ; P. Wu ; Z. Pan ; N. Liu ; X. You (2010) "Dynamic load balancing in 3GPP LTE multi-cell networks with heterogeneous services" 5th International ICST Conference on Communications and Networking in China (CHINACOM), pp. 1 – 5.

[4] Siomina, I. ; Di Yuan, (2012)"Load balancing in heterogeneous LTE: Range optimization via cell offset and load-coupling characterization", in IEEE International Conference on Communications (ICC) ,pp. 1357 – 1361.

[5] Keon-Wook Lee, Jae-Yun Ko, Yong-Hwan Lee,(2006) "Fast Cell Site Selection with Interference Avoidance in Packet Based OFDM Cellular Systems", IEEE GLOBECOM '06,pp.1-5

[6] A. Sang, X. Wang, M. Madihian, and R. Gitlin, (2004)"Coordinated load balancing, handoff/cell-site selection, and scheduling in multi-cell packet data systems," in Proc. ACM Mobicom, pp. 302-314.

[7] T.Qu; D. Xiao; D.Yang,(2010), "A novel cell selection method in heterogeneous LTE-advanced systems " in IEEE 3rd International Conference on Broadband Network and Multimedia Technology (IC-BNMT), Beijing, pp. 510 – 513.

[8] J. Sangiamwong, Y. Saito, N. Miki, T. Abe, S. Nagata, and Y.Okumura, (2011) " Investigation on Cell Selection Methods Associated with Inter-cell Interference Coordination in Heterogeneous Networks for LTE-Advanced Downlink" in IEEE 11th European Wireless Conference, Vienna, pp. 1-6.

[9] Simsek, M.; Hanguang Wu; Bo Zhao; Akbudak, T.; Czylwik, A., (2011), " Performance of Different Cell Selection Modes in 3GPP-LTE Macro-/Femtocell Scenarios" in IEEE Wireless Advanced (WiAd), pp. 126-131.

[10] T. Qu ; D. Xiao ; D. Yang ; W. Jin ; Y. He, (2010)," Cell selection analysis in outdoor heterogeneous networks ", 3rd International Conference on Advanced Computer Theory and Engineering (ICACTE), pp.V5-554 – 557.

[11] RI-083813, "Range expansion for efficient support of heterogeneous networks," Qualcomm Europe.

[12] J. Wang; J. Liu; D. Wang; J. Pang; G. Shen, (2011)," Optimized Fairness Cell Selection for 3GPP LTE-A Macro-Pico HetNets", IEEE Vehicular Technology Conference (VTC Fall), pp. 1-5.

[13] Amzallag, D. Bar-Yehuda, R. Raz, D. Scalosub, (2013),"Cell Selection in 4G Cellular Networks", IEEE Transactions on Mobile Computing, vol. 12, no.7 , pp. 1443 – 1455.

[14] A. Jalali, R. Padovani, and R. Pankaj, (2000), "Data Throughput of CDMAHDR a High Efficiency-High Data Rate Personal Communication Wireless System," in IEEE 51st Vehicular Technology Conference Proceedings, Tokyo, pp. 1854-1858.

[15] C. Westphal,(2004), "Monitoring proportional fairness in cdma2000 high data rate networks," in Proceedings of Globecom, vol. 6, pp. 3866-3871.

[16] B. B. Chen and M. C. Chan,(2006), "Proportional Fairness for Overlapping Cells in Wireless Networks," VTC 2006-Fall, Montreal, Canada, 25-28 September, pp. 1-5.

[17] E. Liu , K. K. Leung, (2008), "Proportional Fair Scheduling: Analytical Insight under Rayleigh Fading Environment", IEEE WCNC ,pp1883-1888.

[18] J. G. Choi and S. Bahk., (2007)," Cell-Throughput Analysis of the Proportional Fair Scheduler in the Single-Cell Environment", IEEE Transactions on Vehicular Technology, , Vol. 56, No. 2, pp. 766 – 778.

[19] X. Qiu and K. Chawla, (1999), "On the Performance of Adaptive Modulation in Cellular Systems," in IEEE Transactions on Communications. vol. 47, 1999, pp. 884-895.

[20] Ramli, H.A.M.; Basukala, R.; Sandrasegaran, K.; Patachaianand, R., (2009), "Performance of Well Known Packet Scheduling Algorithms in the Downlink 3GPP LTE System", IEEE 9th Malaysia International Conference on Communications (MICC) , pp. 815 – 820.

10

A Novel Handoff Decision Algorithm in Call Admission Control Strategy to Utilize the Scarce Spectrum in Wireless Mobile Network

Alagu S[1], Meyyappan T[2]

[1]Research Scholar, Department of Computer Science and Engineering
Alagappa University, Karaikudi, Tamilnadu, India
sivaalagu@hotmail.com
[2]Professor, Department of Computer Science and Engineering
Alagappa University, Karaikudi, TamilNadu, India
meyslotus@yahoo.com

ABSTRACT

Wireless networking is becoming an increasingly important and popular way of providing global information access to users on the move. One of the main challenges for seamless mobility is the availability of simple and robust Channel Allocation algorithm. The radio propagation environment and related handoff challenges are different in cellular structures. A handoff algorithm with fixed parameters cannot perform well in different system environments. In this paper, motivated by the facts that the scarce spectrum should be utilized efficiently, a new Dynamic Channel allocation scheme is proposed. Performance of the proposed dynamic channel allocation scheme is compared with existing channel allocation schemes such as fixed channel, static guard channel and Guard Channel with Channel Borrowing. The experimental results reveal that the proposed algorithm outperforms traditional methods in bandwidth utilization, handoff dropping rate and new call blocking rate.

KEYWORDS

Handoff, Guard Channel, Call Admission, Call Blocking, NHDA, Spectrum

1. INTRODUCTION

Wireless data networks are usually composed of a wired, packet-switched, backbone network and one or more wireless (e.g., cellular radio or infrared) hops connecting mobile hosts to the wired part. The wireless part is organized into geographically-defined cells, with a control point called a *base station* (BS) for each of these cells which is depicted in Figure 1. The base stations are on the wired network and provide a gateway for communication between the wireless infrastructure and the backbone interconnect. As a mobile host (MH) travels between wireless cells, the task of routing data between the wired network and the MH must be transferred to the new cell's base station. This process, known as a *handoff*, must maintain end-to-end connectivity in the dynamically reconfigured network topology.

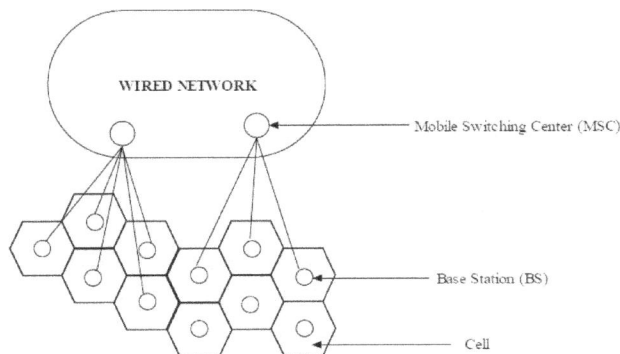

Figure 1. Cellular Architecture

1.1 Mobile Communication System

In modern cellular architecture there are limited available spectrums. The fixed base stations (BS) are interconnected to each other through a fixed network. They communicate with mobile stations (MS) via wireless links. The geographic area is divided into cells in which there is a base station serving each cell. Cells are divided into groups, in which each group is controlled by a mobile switching center (MSC). Neighboring cells overlap with each other to ensure the continuity of communications, when the users move from one cell to another. Fixed number of channels (spectrum) is assigned to each base station.

A channel in the system can be thought of as a fixed frequency bandwidth (FDMA), a specific time-slot within a frame (TDMA), or a particular code (CDMA), depending on the multiple access technique used. BSs and MSCs take the responsibility of allocating channel resources to mobile stations. Same set of channels is reused in another cell far apart enough so that the co-channel interference is negligible. The co-channel reuse distance is defined as the minimum distance at which channels can be reused with negligible interference [1].

1.2 Allowing Mobility of a Subscriber

In order to allow mobility to a subscriber, the cellular network has to have the ability to track down the subscriber when a call is made to them and should allow the subscriber to make calls while they are not in their home BS. Tracking down is only possible if the system maintains information about the location of the subscriber's Mobile device. Also the Mobile device should know the appropriate channels to await signals from the system. In order for both the cellular system and the Mobile device to have this required knowledge, there are two main procedures to follow when the subscriber turns on a Mobile device in any location. They are Searching for channels and Register to the nearest Base Station.

1.2.1 Searching for channels

There are two channels that are searched by the Mobile device which is depicted in Table 1. This procedure enables the Mobile device to identify the correct channels to await signals from the system.

Table 1 .Types of Channels

Channel	Purpose
Strong Dedicated Channel (DCC)	A channel used for the transmission of digital control information from a base station to the Mobile device or vice versa.
Strong Paging Channel	A channel used by the MSC for seeking the Mobile device when a call made to it.

1.2.2. Register to the Nearest Base Station

The Mobile device registers by sending MIN and ESN as shown in Table 2.

Table 2.Access Validation

Access Code	Purpose
Mobile Identification Number (MIN)	The telephone number of the cellular instrument assigned to the subscriber.
Electronic Serial Number (ESN)	This is assigned to the Mobile device by the manufacturer.

The MIN and ESN are used by the MSC for access validation. This involves checking with the information stored in the home base station of the subscriber. Information about the mobile device and its current position is stored in the Mobile Switching Centre (MSC) [2]. Both the information stored in the MSC and the paging channel is used by the MSC to direct any arriving call to the appropriate subscriber. When the subscriber makes calls, they are allowed to be anywhere within the network as this involves the Mobile device accessing base station, which could be located in any of the cells throughout the network. This base station can use the information it has retrieved from the home base station of the subscriber to direct their calls.

During a call, the base station would monitor the signal level from the Mobile device. When the Mobile device is moved into a new cell, the signal level will fall to a critical value causing the base station to inform the MSC about this event. The MSC would instruct the entire surrounding base stations to measure the Mobile device's signal level and transfer the control to the base station receiving the strongest signal level [4]. This is known as hand-over or hand off and occurs within 400 ms. The subscribers are hardly aware of the break in signal. Now registration is done with the new BS. Location information stored in the MSC about this mobile device is updated. If the mobile device is moved into a cell belonging to a different cluster it would also have to register with the new MSC.

2. HANDOFF

Mobile Networks have gained an impulsion in the past few years in rapacious dimensions [3]. And since then mobility becomes a distinct feature of wireless mobile cellular system [5]. While a call (mobile caller/user in service) is in progress the channel (frequency, time slot, spreading code, or combination of them) associated with the current connection is changed through Channel Allocation Control (CAC) proposals [6]. The existing call may change its present Base Station (BS) also termed as Mobile Terminal (MT) to a new one. This phenomenon is whatever we call handover (handoff). It is shown in Figure 2.

Usually, this handover mechanism supports continuous services by transfer of an ongoing call from the current cell to the next adjacent cell as the mobile (MS) moves through the coverage area. Either crossing a cell boundary of current BS by mobile station (MS) or deterioration in quality of the signal in the current channel is the primary responsible factor for initiating a handover [6][5].

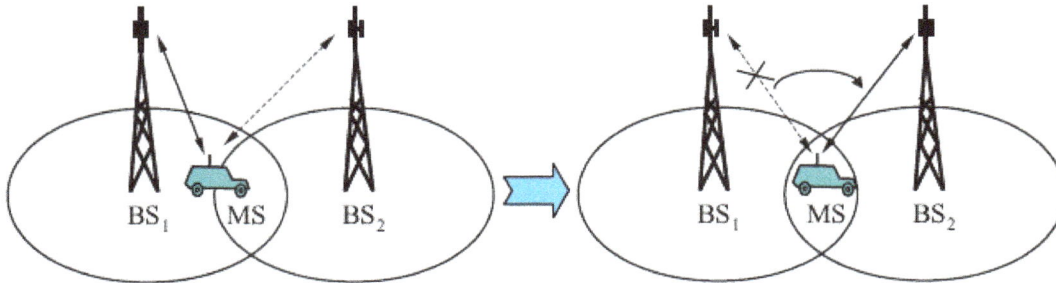

Figure 2. Handoff Scenario

A successful handoff provides continuation of the call which is vital for the perceived quality of service (QoS). In case the next cell does not have a radio channel available for the incoming MS, handoff blocking occurs and the call is dropped. The lack of channel resources also results in the blocking of new calls. The universally accepted design concept in cellular networks is that blocking of handoff requests is less desirable than the blocking of new calls. The QoS is mainly determined by the two blocking probabilities and the overall resource utilities. One of the important objectives in the development of the new generation is improving the quality of cellular service, with handoffs nearly invisible to the MSs.

Network protocols in cellular wireless data networks must update routes as a mobile host moves between cells. Most current handoff schemes in wireless networks result in data loss or large variations in packet delivery times. Unfortunately, many applications, such as real-time multimedia applications and reliable transport protocols, adapt to long term estimates of end-to-end delay and loss. Violations of these estimates caused by handoff processing often result in degraded performance. For example, loss during handoff adversely affects TCP performance and high packet loss and variable delays result in poor real-time performance.

2.1 Channel Allocation

In wireless mobile networks, the service area is divided into cells each of which is equipped with a number of channels. New originating calls in the cell coverage area and the handoff calls are sharing these channels. When any of these calls arrives at a cell where channel is not available, it has to be blocked or queued or rejected depending on the call admission control schemes. The probability of the new originating call in the cell that is rejected is called Call Rejection probability and the probability that a handoff call rejected is called Handoff Rejection Probability.

Generally the Handoff request is initiated either by the Mobile Station or by the Base Station. Different types of handoff decision protocols are used in various cellular systems. Some of them are:

- **Network Controlled Handoff**

  In this scheme, the Mobile Switching Center (MSC) is responsible for the overall handoff decision [1]. MSC measures the signal strength and receivers threshold from different Base Stations. It then decides on the handoff request to a Base Station whose signal level is closest [1][14].

- **Mobile Assisted Handoff**

  Here the Mobile Station (MS) is responsible for finding the Base Station (BS) whose signal strength is closest to it. The MS measures the signal strengths periodically in the neighboring BS. Based on the received measurements, the BSs and MSC decides when to handoff [9][14].

- **Mobile controlled Handoff**

  In this scheme the MS has got the full control in handoff decision. Both BS and MS measures the signal strength in the neighboring BSs and the current BS sends them to the MS. The MS decides when to handoff based on the information gained from the BS and itself [9][10][14].

2.2 Performance Metrics for Handoffs

The following are considered as the performance metrics for handover of calls from one cell coverage area to the others.

- *Call blocking probability:* The probability that a new call attempt is blocked.
- *Handoff blocking probability:* The probability that a handoff attempt is blocked.
- *Handoff probability:* The probability that while communicating with a particular cell, an ongoing call requires a handoff before the call terminates. This metric is translated to obtain the average number of handoffs per cell.
- *Call dropping probability:* The probability that a call terminates due to handoff failure. This metric can be derived directly from the handoff blocking probability and the handoff probability.
- *Rate of handoff:* The number of handoff per unit time.
- *Duration of interruption:* The length of time during handoff for which the mobile terminal is in communication with neither base station.
- *Delay:* The distance that the mobile user moves from the point at which, the handoff should occur to the point at which it does.

3. LITERATURE SURVEY

Many papers in the literature of related work addresses the categorization of the schemes Based on Guard channel concept. In the cellular network, channel assignment strategies can be classified into fixed, flexible and dynamic [7].The existing literature addresses the Static Guard Channel allocation exclusively for handoff and fixed channel system where there are no separate guard channels exclusively for handoff. In fixed channel assignment (FCA) scheme, fixed numbers of channels are assigned to each cell and there isn't any Guard Channel set aside exclusively for handoff requests. Whenever new call request or handoff request arrives, the base station will check to see if there is a channel available in current cell. The call will be connected if there is a channel available and it will be dropped if there isn't any channel left. So handoff request and new call request are dealt with equally. The cell doesn't consider the difference between Handoff request and new call request. It assigns the channels to BS by First Come First Serve basis [5][12][13]. The Quality of Service is not satisfied because the handoff blocking rate is as same as new call blocking rate. The so called "Guard-channel" (GC) concept offers a means of improving the probability of a successful handoff by reserving a certain number of channels allocated exclusively for handoff requests. The remaining channels can be shared equally between handoff requests and new calls [1][9]. Allocating Guard channels for Handoff improves the overall throughput which was discussed in our previous papers [12][13]. If the guard channel number is too big, the new call blocking rate will be high because several channels are set aside for handoff requests even when the traffic load is low. In this case, the resources are wasted by not serving either for handoff request or new call request. If the number is too small, the handoff blocking rate can't be guaranteed under high traffic load. So this scheme enhances the QoS by reducing the handoff blocking rate in a stable traffic load. While when the traffic load is changing periodically or dynamically due to big event or working rush hours, it is not flexible enough to get good QoS. This scheme sometimes ends up with inefficient spectrum utilization. Careful estimation of channel occupancy time distributions is essential in order to minimize this risk by determining the optimum number of guard channels. Static Guard Channel Allocation with Channel Borrowing scheme [15] does improves the channel utilization but the complexity of the algorithm increases since preemption is done in every allocation.

4. PROPOSED WORK

In this paper, the authors devise a scheme – A Novel Handoff Decision Algorithm in Call Admission Control Strategy (NHDA). In this proposed new scheme, the channels for handoff requests are dynamically allocated based on the handoff failure probability observed for a certain past period in the network. This scheme aims to utilize the scarce spectrum efficiently and also to balance the load in the network traffic. The new call dropping rate determines the fraction of new calls that are rejected. The handoff blocking rate is closely related to the fraction of admitted calls that terminate prematurely due to handoff. Limited channels, a scarce resource, should be utilized effectively. Efficient resource utilization is the main objective of this research work. For effective resource utilization, less number of Guard channels should be assigned to the handoff calls during low traffic load in the network. If more channels are reserved for the handoff request in this condition, the resources are wasted as the channels serve neither for handoff request nor a new call request. On the other hand, if the number of handoff request is more than the number of available Guard channels, then the number of guard channels should be increased. The balance of the new call rejection rate and handoff call rejection rate are monitored and maintained to get better resource utilization in cellular network. The Channel allocation model is shown in Figure 3.

A call being forced to terminate during the service is more annoying than a call being blocked at its start. Hence the handoff call blocking probability is much more stringent than new call blocking probability. Therefore it is intuitively clear that priority is given to handoff requests by assigning **GCh** channels exclusively for handoff calls among the C channels in a cell. The remaining **Oc (=C-GCh)** channels are shared by both originating calls and handoff requests. The selection of number of guard channels exclusively for handoff call is essentially important factor to get good Quality of Service. In the proposed scheme the guard channel **GCh** is initially assigned and dynamically altered based on the traffic in the network.

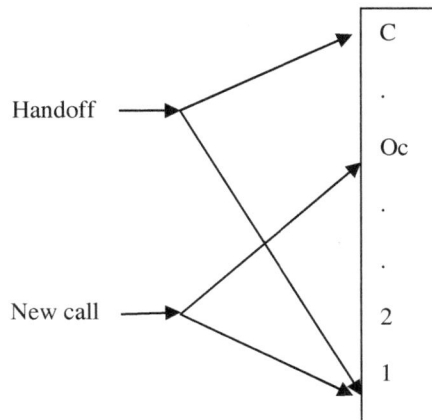

Figure 3. Channel allocation model with priority for handoff calls

4.1 Call Admission Control

Quality of Service (QoS) provisioning in wireless networks is a challenging problem due to the scarcity of wireless resources, i.e. radio channels, and the mobility of users. Call Admission Control (CAC) is a fundamental mechanism used for QoS provisioning in a network. It restricts the access to the network based on resource availability in order to prevent network congestion and service degradation for already supported users. A new call request is accepted if there are enough idle resources to meet the QoS requirements of the new call without violating the QoS for already accepted calls. With respect to the layered network architecture, different QoS parameters are involved at different layers. At physical layer, bit-level QoS parameters such as bit energy-to-noise density describe the quality of service [3]. In packet-based communication systems, packet-level QoS parameters such as packet loss, delay and jitter characterize the perceived quality of service [3]. However, most of the existing research on call admission control in cellular networks has focused on an abstract representation of the network in which only call-level QoS parameters, namely, call blocking and dropping probabilities are considered. When a mobile terminal (mobile user) requests service, it may either be granted or denied service. This denial of service is known as call blocking, and its probability as *call blocking probability* (Pb). An active terminal in a cellular network may move from one cell to another. The continuity of service to the mobile terminal in the new cell requires a successful handoff from the previous cell to the new cell. A handoff is successful if the required resources are available and allocated for the mobile terminal. The probability of a handoff failure is called *handoff failure probability* (Pf). During the life of a call, a mobile user may cross several cell boundaries and hence may require several successful handoffs. Failure to get a successful handoff at any cell in the path forces the network to discontinue service to the user. This is known as call dropping or forced termination of the call and the probability of such an event are known as *call dropping probability* (Pd). In general, dropping a call in progress is considered to have a more negative impact from the user's perspective than blocking a newly requested call.

According to the above definition, the call dropping probability, Pd, and handoff failure probability, Pf, are different parameters. While the handoff failure probability is an important parameter for network management, the probability of call dropping (forced termination) may be more relevant to mobile users and service providers. Despite this fact, most research papers focus on the handoff failure probability because calculating Pf is more convenient.

4.2 Channel Assignment Schemes

Channels are managed at each cell by channel assignment schemes based on co-channel reuse constraints [6]. In Static Guard Channel Allocation scheme, a set of channels is permanently assigned to each base station. A new call can only be served if there is a free channel available in the cell. Due to non-uniform traffic distribution among cells, Static Guard Channel Allocation schemes suffer from low channel utilization. Channel borrowing scheme overcomes this problem at the expense of increased complexity and signaling overhead. In this research work the authors have proposed the Novel Handoff Decision Algorithm in Call Admission Control Strategy (NHDA) where the channels are allocated dynamically based on the observation in the traffic.

4.3 Novel Handoff Decision Algorithm (NHDA)

The selection of number of guard channels exclusively for handoff call is essentially important factor to get good Quality of Service. For different type of traffic load and mobility factor, different number of guard channels is needed to be allocated. The number of guard channels can't be fixed when the traffic load is changing with the time. The authors address this problem through the proposed scheme NHDA.

The proposed scheme automatically searches the optimal number of Guard Channels to be reserved for handoff calls at each BS. In this paper the author considers for a Base Station BS, having total number of channels C, the Guard channels exclusively for handoff will be GCh. The rest of the available channels are used by the new originating calls in that cell and also by the handoff calls, which is say Co. A new call request will be granted for admission if the total number of on-going calls (including handoff calls from other cells) is less than the number Co. A handoff call request will be granted for admission if the total number of on-going calls in the cell is less than the total capacity C.

The algorithm NHDA can be illustrated as follows:

Data Structure

Consider the following parameters in a particular BS coverage area,
The total number of available channels   – C
Open Access Channels (new calls + Handoff calls) – Co
Guard channels for handoff calls   – GCh
Where, C = Co + GCh, Co = C – Gch and GCh is allocated dynamically
Oc = number of on-going calls
Nc = number of admitted new originating calls
Hc = number of admitted handoff calls
H=Total number of handoff call (admitted+rejected)

Where, Oc = Nc + Hc
Pd = Call dropping probability // used in FCA scheme
Pf = Probability of Handoff failure
Pb = Call blocking probability
t = time period
Th = Threshold for handoff call rejection probability

Algorithm: NHDA (t, C) // the algorithm takes time period and channels as input
{
Co=C-GCh
For every handoff call request Do
{
   If Oc < C, then
     {
       Hc = Hc + 1 and grant admission
       Oc = Oc +1
     }
Otherwise, Pf = Pf +1 and reject.
}

For every new call request Do
{
   If Oc < Co, then
   {
   Nc=Nc+1 and grant admission
   Oc = Oc +1
   }
   Otherwise, Pd = Pd +1 and reject.
}
If a call is completed or handoff to another cell
{

Oc = Oc – 1
Check with MSC whether the ended call is handoff call or new originated call
If handoff call then Hc = Hc-1
Else Nc = Nc-1
}

If a handoff call is dropped and Pf/H >= AuTh then
{
GCh = min {GCh +1, Cmax}

If Pf/H <= AdTh for N consecutive handoff calls, then
GCh = max {GCh – 1, Cmin}
}
Nc and Hc are reported to understand the successful handoff and new calls at a specified time period.

}//end of the algorithm: NHDA

As the number of Guard Channels allotted plays a vital role to the key performance, it is dynamically altered every specific time period say t. In this approach the number of guard channels which is to be allocated is determined through optimizing certain performance goal with service quality constraints. When a base station experiences high handoff blocking rate, the number of guard channels will be increased until the handoff blocking rate drops to below its threshold. When a base station does not get to use a significant portion of the guard channels over a period of time, the number of guard channels is gradually decreased until most of the guard channels are used frequently. By doing this, the handoff blocking rate is controlled to close to its threshold.

The proposed algorithm increases the number of guard channels GCh, when a handoff call is dropped under the condition that Pf/H >= Au*Th, and it decreases the number of guard channels after a number of consecutive handoff calls under the condition that Pf/H <= Ad*Th. Au and Ad are usually chosen to be less than 1. By choosing Au < 1, the algorithm will most likely keep the handoff blocking rate below its given threshold.

The simulation studies are performed for comparisons of the proposed algorithm with fixed channel allocation (FCA) and static guard channel allocation policy. The result proves that the algorithm guarantees that the handoff dropping rate is below its given threshold and at the same time the new call dropping rate is minimized.

5. EXPERIMENTAL RESULTS

In the simulation study of the NHDA, the authors used a model that adhered to the general assumptions made in the literature. Below, the authors describe the various components of the simulation model and the assumptions for these components.

5.1 Cell Model

In the simulation, the authors use 2-D cellular system model with wrap around which is depicted in figure 4. Our simulation tests use a 6*6 cellular patch with wrap around, a cell radius of 1000m, a minimum reuse distance of 3, and a TD equal to 0.8 of the cell radius. MSs are allowed to wrap around to the other side of the system when moves out of system boundary. It eliminates the burden of handling out of bound situations and is considered an efficient way to approximate the simulation of a very large cellular system. Each cell is considered as a circle and has exactly six neighbors. Object oriented approach is used to implement the real world environment. Figure 4 shows the cell structure considered for simulation.

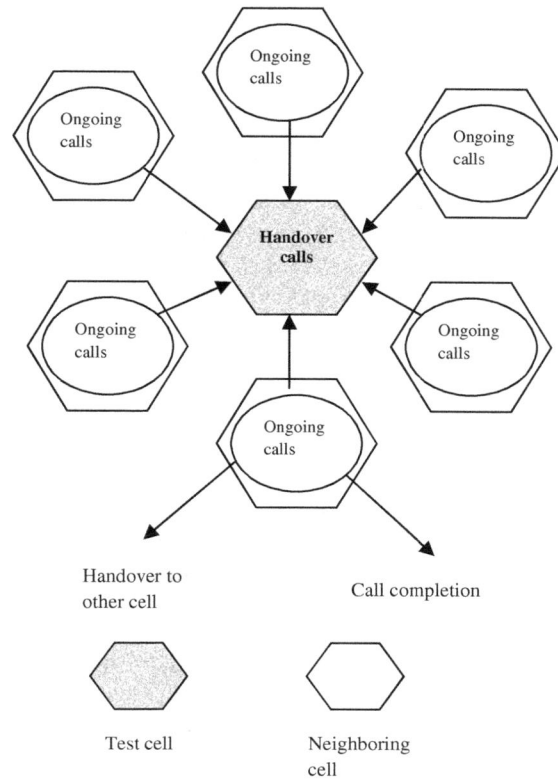

Figure 4. Simulation model for a single Cell

5.2 Traffic Model

The authors use exponential distribution to determine the duration of each generated call with a mean of 180s. New Calls arrive according to a Poisson process and homogeneous traffic among all cell were considered. Each cell is assigned 20 channels. The traffic load to each cell is defined as:

$$\frac{Arrival\ Rate\ to\ the\ cell * Average\ call\ duration}{Number\ of\ channels\ per\ cell} * 100$$

5.3 Mobility Model

In our model, each MS is assigned initial speed and direction with an average speed of 18 meters/s and a maximum speed of 24 meters/s. After a specified time period, which is generated randomly, the speed and direction of the MS are updated.

The selection of ideal time period is an important parameter too. If the time period is too small, the overhead in network is too high and the algorithm can't take much advantage. If the time period is too large, the handoff blocking rate will be adjusted to fit to the changing traffic load too slowly and as a result the QoS will not be guaranteed. The authors have chosen 2 hours time period to test and simulate the scenario.

5.4 Simulation results

Extensive simulations are conducted to evaluate the performance of NHDA scheme.

The result comprises of comparison between the 4 schemes as Handoff Handling without using Guard Channels-FCA, Using static Guard Channels, using Guard Channel with Channel borrowing scheme, and the proposed scheme NHDA. The parameters chosen for simulation are:

busy_channels   : Number of channels occupied by calls.
next_event_type : Type of next event New call, New Handoff, Channel release.
total_calls   : Number of calls generated in or handed to the cell.
new_success   : New calls which have been assigned channels by the BS.
ho_success   : Handovers which have been assigned channels by the BS
ho_fail   : Handovers which have not been assigned channels.
Blocked   : New calls which have not been allocated channels.
incell_success   : New incell call or handover which have been assigned channels.
incell_blocked   : New incell call or handover which have not been assigned channels.
incluster_success : New incluster call or handover which have been assigned channels.
incluster_blocked: New incluster call or handover which have not been assigned channels.
outcluster_success: New outcluster call or handover which have been assigned channels.
outcluster_blocked: New outcluster call or handover which have not been assigned channels.
call_type   : Type of call; Incell, Incluster, Outcluster
BTS_index   : Index of Base Station whose event will occur.
Capacity   : Load which a backbone link can handle.
next_call   : Time at which next new call will be generated.
next_event_time : Time at which next event will occur.
next_handover   : Time at which next new handover will be generated.
ho_delay   : Time for which a handover is stored in the handover queue.
miat   : Mean inter arrival time. Time difference between successive calls
hmiat   : Handoff Mean inter arrival time. Time difference between successive handover.

Oc=0, Nc=0, Hc=0, Pb=0, Pf=0,H=0
t=20 seconds // time period for updating the measurements
C=20// No. of channels
GCh=10 // Guard Channels exclusively for Handoffs
Au=0.9
An=0.6
Th=0.8
N=10 //No. of consecutive calls

Number of cells: 36
Position measurement interval: 3 sec.
Mean call duration: 180 sec
Average speed of an Ms: 18 m/s
Simulation time: 200,000 sec.

Number of Radio Channels = 50 per cell.
Number of Local Link Channels = 50 per cell.
Average time for a new call = 60 sec.
Average time for a handover call = 45 sec.
Maximum handover queue time = 15 sec.
Capacity of Backbone Links = 75 calls.

Formulae Involved:

Call Blocking   = $\dfrac{\text{Total number of calls blocked}}{\text{Total number of calls processed}}$

Handoff failures =
$$\frac{\text{Total no. of handoffs not assigned Channels}}{\text{Total number of calls processed.}}$$

Throughput   =
$$\frac{\text{TSC + TSH}}{\text{Total number of calls processed}}$$

Where,
TSC = Total number of calls that have been assigned channels and backbone links.
TSH=Total number of handoffs that have been assigned channels and backbone links.

The following graphs show the comparative study of the four schemes, Fixed Channel Assignment without using Guard Channels, Static Guard channel assignment, Guard Channel assignment with Channel borrowing scheme and the proposed scheme NHDA.

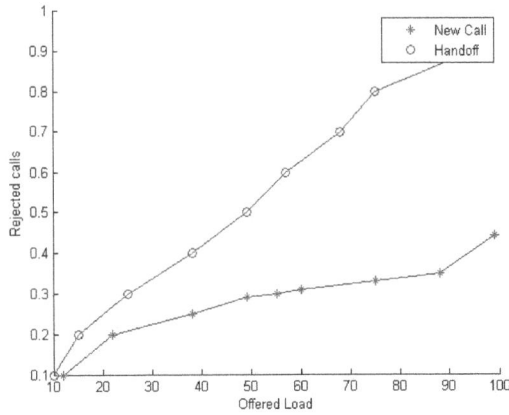

Figure 5. New Call handling and Handoff call handling without using Guard Channels (FCA)

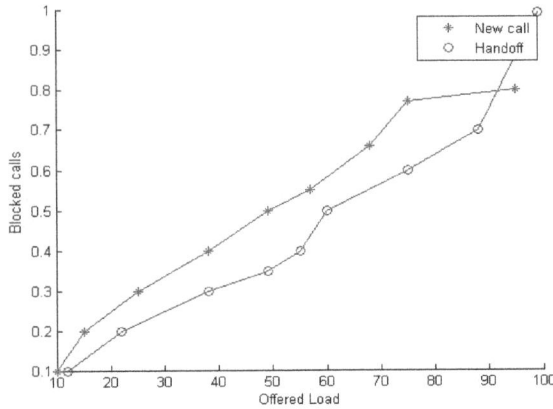

Figure 6. Static allocation of Guard Channels exclusively for Handoffs

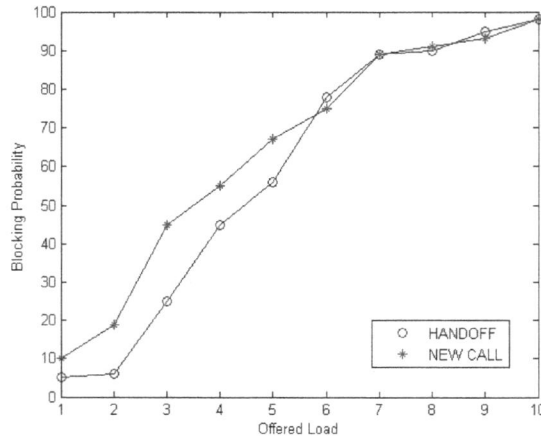

Figure 7. Guard Channel Allocation with Channel Borrowing Scheme

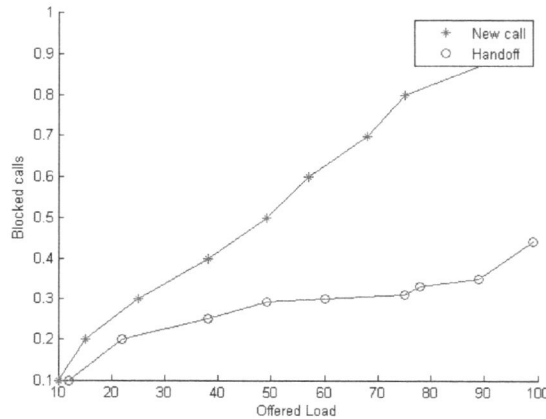

Figure 8. Novel Handoff Decision scheme

Figure 5 shows the simulated output of the FCA scheme where no guard channels are allocated for handoff calls. It simply works in FCFS (First Come First Served) manner. The output graph shows that the Handoff Call rejection rate is slightly high than the new originating call rejection rate. But it should be noted here that handoff calls should be given higher priority than the new originating calls. The overall performance is also not satisfactory since both new call rejection and handoff call rejection ratio is comparatively high.

Figure 6 shows the simulated output of the static guard channel allocation scheme i.e., fixed number of guard channels exclusively allocated for handoff. Here the number of handoff call rejection is reduced but the new call rejection is highly increased because the number of guard channels allocated is high than which is actually required. Moreover in some cases if the number of guard channels is less, then handoff rejection rate will increase and hence affect the throughput.

Figure 7 shows the simulated output of the Guard Channel Allocation with Channel Borrowing scheme. The graph shows that the handoff blocking and new call blocking varies often as borrowing strategy is implemented. Here the handoff blocking rate is reduced but the complexity of the algorithm increases and as a result of that the overall throughput is not satisfactory.

Figure 8 shows the simulated output of our proposed scheme – Novel Handoff Decision Algorithm (NHDA). Here as the channels are not allocated static and they are allocated based on the traffic in the past certain period of time. The number of guard channels gets dynamically adjusted and it is clearly seen from the graph that both new calls and handoff calls utilizes the channel properly and the call rejection rate is low for both. Hence there is tradeoff.

6. CONCLUSION

This paper presents a new algorithm for call admission control in a mobile cellular network. From the graph it is obvious that the new strategy shows better resource utilization.

Due to the unique characteristics of mobile cellular networks, mainly mobility and limited resources, the wireless resource management problem has received tremendous attention. As a result, a large body of work has been done extending earlier work in fixed channel allocation by introducing new techniques. In this paper, the authors presented a Novel Handoff Decision Algorithm in call admission control strategy for cellular networks. This new scheme utilizes the available resources effectively. In this work, a significant contribution has been made in the area of call admission control with the hope of improving the Call Admission Control performance.

For evaluating the performance of this scheme, a simulation model was created and lots of tests were done. The simulation result shows that the proposed algorithm can adapt to the changes in traffic conditions (changes in the call arrival rate) and can achieve optimal performance in terms of guaranteeing handoff call blocking threshold and at the same time minimizing the new call blocking rate. This adaptive approach can automatically search the optimal number of guard channels to be reserved at a base station. Existing Guard channel allocation schemes lack dynamism to cope up with dynamic network traffic. The proposed algorithm adjusts the number of guard channels dynamically according to the dropping rate of handoff calls in a certain period of time. It either increases or decreases the number of guard channels allocated based on observed handoff rejection threshold. The proposed scheme NHDA possesses high degree of spectrum utilization with good QoS and is a simple algorithm with a satisfied implementation complexity.

REFERENCES

[1] Yi Zhang and Derong Liu, *"An Adaptive Algorithm for Call Admission Control in Wireless Networks"*, IEEE 2001

[2] Qing-an-Zeng and Dharma P.Agarwal, *"Handbook of Wireless Networks and Mobile Computing"*,John wiley & sons, Inc., 2002

[3] Kishore S Trivedi et.al *"Analytical Modeling of Handoffs in Wireless cellular networks"*, Elsevier Information Sciences, 2002

[4] A. Iera, A. Molinaro and S. Marano, *"Handoff Management with Mobility Estimation in Hierarchical Systems"*, IEEE Transactions on Vehicular Technology, vol. 51, Sept. 2002.

[5] I.Ramani and S.Savage. SyncScan *"Practical fast handoff for 802.11 Infrastructure Networks"*, Proceedings of IEEE INFOCOM, March 2005

[6] Nasif Ekiz, Tara Salih, Sibel Kucukoner and Kemal Fidanboylu, *"An overview of Handoff techniques in Cellular Networks"*, World Academy of Science, Engineering and Technology, 6 2005

[7] F. Siddiqui and S. Zeadally, *"Mobility Management across Hybrid Wireless Networks: Trends and Challenges,"* Computer Communications, vol. 29, no. 9, pp. 1363–1385, May 2006.

[8] Nasser, N., Hasswa, A. and Hassanein, *"Handoffs in Fourth Generation Heterogeneous Networks"*, IEEE Communications Magazine, vol. 44, pp. 96-103, 2006.

[9]   Rami Tawil, Jaques Demergian, Guy Pujolle, *"A Trusted Handoff Decision Scheme for the Next Generation Wireless Networks"*, IJCSNS International Journal of Computer Science and Network Security, VOL.8 No.6, June 2008

[10] Yoo, S.-J.,et al., *"Analysis of fast handover mechanisms for hierarchical mobile IPv6 network mobility"*, Wireless Personal Communications, 48(2), 215-238., 2009.

[11] Asd Malak Z. Habeib, Hussein A. Elsayed, Salwa H. Elramly and Magdy M. Ibrahim, *"Heterogenous Networks Handover Decision Triggering Algorithm Based on Measurements Messages Transfer using IP Option Header"*, The Seventh International Conference on Wireless and Mobile Communications - ICWMC 2011

[12] Alagu.S and Meyyappan.T, *"Analysis of Algorithms for Handling Handoffs in Wireless Mobile Networks"*, International Journal of P2P Network Trends and Technology- Vol1Issue2, 2011.

[13] Alagu S and Meyyappan T, *"Analysis of Handoff Schemes in Wireless Mobile Network"*, IJCES International Journal of Computer Engineering Science, Vol1 Issue2, Nov. 2011.

[14] V. S. Kolate, G. I. Patil, A. S. Bhide, *"Call Admission Control Schemes and Handoff Prioritization in 3G Wireless Mobile Networks"*, International Journal of Engineering and Innovative Technology (IJEIT) Vol1, Issue3, March 2012

[15] Alagu.S and Meyyappan.T, *"Guard Channel Allocation with Channel Borrowing Scheme to handle Handoffs in Wireless Mobile Networks"*, Proceedings of the National Conference on Computer Applications, Bharathiar University, March 2012.

11

Context Aware Resource Allocation in Distributed Sensor Networks

Lokesh. B. Bhajantri[1], Nalini. N[2], Gangadharaiah. S[3]

[1]Department of Information Science and Engineering,
Basaveshwar Engineering College, Bagalkot, Karnataka, India.
lokeshcse@yahoo.co.in

[2]Department of Computer Science and Engineering,
Nitte Meenakashi Institute of Technological, Yelahanka, Bangalore. Karnataka, India.
nalinaniranjan@hotmail.com

[3]Department of Information Science and Engineering,
Acharya Institute of Technology, Soladevanahalli, Bangalore.
gangadhar.s@gmail.com

ABSTRACT

Distributed Sensor Networks (DS'.s) have been an area for active research over the past few years, due to potentially widespread applications in creating smart computing environment. In this work, we have proposed new smart computing environment for super/hyper market. The proposed system uses Distributed Sensor Networks (DSN's) for real-time tracking of the products. While new technologies are making sensors nodes smarter, smaller, and cheaper, it is very challenging to find an optimal to allocate sensors and other network resources.

Context awareness is one of the methods to improve resource utilization in Distributed Sensor Networks (DSN's). We have studied Indian retail industry as case study and we listed out the drawbacks of Indian retail industry. But proposed system can be applied to most of the retail industries across the globe to overcome similar problems. As a part of proposed system, we have derived many contexts in retail industry based on human perspective. At a basic level of design, sensors are deployed in each rack which provides the real time information about available products in each rack. This information can be used by warehouse manager, store manager for day to day activity in super/hyper market. R&D team can also use this information to understand consumer behaviour. The context information derived from sensor node is used in allocation sensor node or resources. We have tested our proposed system using Castalia simulation tool, a derivative of OMNeT++. The results obtained from our experiment shows advantages of context aware system. As DSN's are data centric, analyzing and implementing system in user context prospective will greatly contribute in terms of cost, performance. We have simulated the context aware resource allocation in distributed sensor networks to test the operation scheme in terms of performance parameters. The DSN's promise to revolutionize sensing in a wide range of application domains. This is because of their reliability, accuracy, flexibility, cost effectiveness and ease of deployment, Smart sensors can offer vigilant surveillance and can detect and collect data.

KEYWORDS

Distributed Sensor Networks (DSN's), Context, Retail Industry, Store Manager, Floor Manager, Cluster Head, and OMNET++.

1. INTRODUCTION

Advances in hardware and wireless network technologies have created low-cost, low-power, multifunctional miniature sensor devices. These devices make up hundreds or thousands of ad hoc tiny sensor nodes spread across a geographical area. These sensor nodes collaborate among themselves to establish a sensing network. A Distributed Sensor Network (DSN) that can provide access to information anytime, anywhere by collecting, processing, analyzing and disseminating data. Thus, the network actively participates in creating a smart environment. Distributed Sensor Networks (DSN's) have been one of the choices for building smart environments. Smart environments can be designed for building, utilities, military, shopping mall, industrial, home, shipboard, and transportation systems automation. Sensory data generated from multiple sensors in the network is used for creating smarter environment. The main goal of DSN's is to make decisions or gain knowledge based on the information fused from distributed sensor inputs. At the lowest level, individual sensor node collects data from different sensing modalities on-board. An initial data processing can be carried out at the local node to generate local event detection result. These intermediate results will then be integrated/fused at an upper processing center to derive knowledge and help making decisions. A general DSN consists of a set of sensor nodes, a set of processing elements (PE's), and a communication network interconnecting the various PE's [1, 2].

While this new kind of networks has wide range of applications, it also poses serious challenges like frequent network topology change, limited computational, memory and power supply [3, 4]. Sensors are more prone to failures. With all these constraints an efficient and effective method to extract data from the network is challenging task. To address these challenges DSN have flexible design both in hardware, software. Hardware for sensor node is designed with dynamic voltage scaling, different level of sleep modes, hardware level data fusion & aggression, intelligent signal processing, small battery, remote configuration etc. In software, the operating system is designed to work under constrained memory, computation power, battery power, with limited bandwidth etc. The communication system is also designed to achieve greater energy efficiency. Different protocols like SMAC, LEACH, PEGASIS, LSU and TEE are designed specially for energy efficiency [4, 5]. The benefits and limitation of DSN are: DSN provide smart method of computation. The data gathered from different sensors can be utilized to derive more meaningful context. But it also posses many challenges like node should be cost effective, energy efficient, smaller size, stable and standard middleware. The network developed from this sensor should be self configuring and fault tolerant, application specific and data centric.

Context awareness has great potential for creating new service modes, resource allocation and improving service quality in distributed sensor networks. The various types of contexts in distributed sensor networks are: computing context, user context, physical context, location context and time context. Context could be active or passive. In active context awareness, an application automatically adapts to discovered context, by changing the application's behaviour and in passive context awareness, an application presents the new or updated context to an interested user or makes the context persistent for the user to retrieve later [6]. Context aware computing extends the idea of contexts to refer the physical and social situations, in which computational devices are located.

In this paper, we have considered 3 types of contexts are: Time, seasonal and hybrid context in distributed sensor network environments for resource allocation. Context means situational information or can be stated as: "Context is any information that can be used to characterize the situation of an entity". Context aware computing aims to provide maximal flexibility of a computational service based on real time sensing of any forms of contexts.

Context awareness has great potential for creating new service modes, resource allocation and improving service quality in distributed sensor networks. The various types of contexts considered in distributed sensor networks are: computing context, user context, physical context, location context and time context. Context could be active or passive. In active context awareness, an application automatically adapts to discovered context, by changing the application's behaviour and in passive context awareness, an application presents the new or updated context to an interested user or makes the context persistent for the user to retrieve later. Context aware computing extends the idea of contexts to refer the physical and social situations, in which computational devices are located [7].

OMNeT++ is a public-source, discrete event simulation, component-based, modular and open-architecture simulation environment with strong GUI support and an embeddable simulation kernel. Its primary application area is the simulation of communication networks. Because of its generic and flexible architecture it has been successfully used in other areas like the simulation of IT systems, queuing networks, hardware architectures, and business processes. OMNeT++ is rapidly becoming a popular simulation platform in the scientific community as well as in industrial settings. Several open source simulation models have been published, in the field of Internet simulations (IP, IPv6, MPLS, etc), mobility and ad-hoc simulations and other areas. However, such a growing community faces also growing challenges and problems: Integration of different simulation tools, porting of simulation models between different platforms, testing and comparison of applications [8].

The rest of the paper is organized as follows. Section 2 describes the problem overview. Section 3 discusses related work in this area. Section 4 describes the proposed work. Section 5 presents the simulation procedure, performance parameters and the results of the proposed work. Finally, we conclude in section 6.

2. PROBLEM OVERVIEW

The Indian retail industry is moving steadily from unorganized sectors to organized sector. With this huge growth rate, organized retail industries are facing many challenges, mainly proper IT infrastructure to manage the day to day activities in all its outlets. So there is an acute need for smarter computing devices to reduce the burden on retail chains. Other challenges faced by Indian retail industry are absence of developed supply chain and integrated IT management, rising rental & labour values, mall management, shortage of trained staff, manual resource allocations, lack of quality locations, regulations restricting real estate purchases, and cumbersome local laws, and taxation, which favours small retail businesses.

2.1 Present Working Model of Indian Retail Industry

The Figure1 explains the present working model of Indian Retail Industry. At the end of the day, store manager get the transaction summary about sold items. This transaction summary is system generated and will be sent to warehouse before they close the outlet shop. Next day the warehouse division will analyze the system generated file and plan for the further processing. All these process will take at least one day before product is replaced in each outlet. In some retail shops, we have observed that this process is taking more than two days. So there is always a chance that outlet may not have item what consumer demands for which result in loss of business.

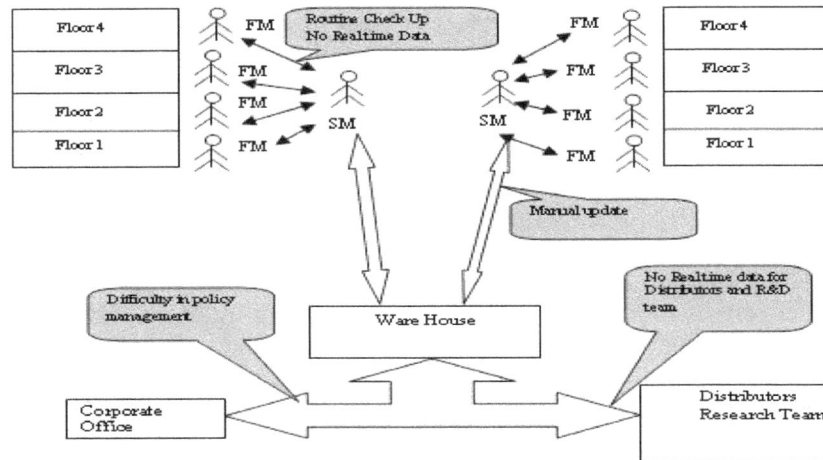

Figure1. Present working model of Retail Stores.

This process model has certain setbacks: 1) Floor manager has to check manually each part of the floor to track the item sold 2) During major festivals or weekend products may sold earlier than expected 3) Always Store Manager (SM) should be in contact with each Floor Manager (FM) for product availability in each floor 4) Some unexpected events like marriage, meeting or some other type of events may happen in around the outlet shop. In this case product may sell earlier than expected by Store manager 5) It increase the labour cost as each floor manager will has to check by himself or with help of other 6) Since all the decision are taken manually, there always chance for human error factor, 7) No visibility to warehouse division about present transactions, 8) Outlet stores have limited space as most of them inside the city. Especially in case of India where real estate cost is considerably high.

In our proposed system, DSN's are used to track the products on each rack of Retail shop. We have developed model by considering only the weight sensors. But depending on type of the products to be tracked, DSN's may be designed to have one or more different type of sensors. We also derived many context related to retail industry. These contexts are used for smarter resource allocation and management. Main contexts derived in our system are: Seasonal Context, Time Context and Hybrid Contexts.

In the proposed system, each sensor node is pre-configured to fixed weight and Context (Either Seasonal or Time context base). After initial setup, each sensor nodes will sample data in a regular interval. Depending on the context and available information, each node will derive an emergency value which signifies the attention to be given to that particular tray. Derived value is forwarded to the Cluster Head (CH), which summarize further with other sensor nodes. Sensor nodes will also modify the event sensing rate or sleep time depending on available information, to reduce the power consumption. Catering the service to every floor by Store manager (SM) solely depends on emergency value forwarded by DSN. The cluster head (CH) acts as intermediate node to forward the data to SINK node. It collects the data from all the node belongs to its cluster. Then it summarizes the data, mainly the emergency value of particular product. Then it forward the data to next cluster which connect to SINK or another cluster.

3. RELATED WORK

We have not found any work related to real time product tracking in retail industry with context aware DSN's for resource allocation. We derived many contexts, related to retail industry. These contexts are used in product tracking and resource allocation. In this section discuss about related work in context aware systems and resource allocation. Creating meaningful knowledge from raw data gathered from sensor challenging task.

The work given in [9] presents a context aware approach to conserving energy in wireless sensor networks. As sensor hardware technology proliferates, research in prolonging sensor battery life gains more interest. Conservation of sensor energy, therefore, becomes a practical approach to prolonging sensor life and eventually reducing the frequency of battery replacement. In retrospect, hardware driven approaches have enabled significant power savings but on the expense of genericity as they are typical to certain sensor hardware. In this paper, they proposed a software based approach that is not constrained by sensor hardware and which is independent of any specific application domain. An implementation is done utilizing the underlying framework.

The work given in [10] discusses many applications of context-aware system including the context aware pill bottle for elderly persons. The system remind elderly users when it is time to take their medication, and a medication monitor situated in a caregivers home that displays awareness information for elder users medication compliance. Context-aware system provides countless opportunities.

The work given in [11] discusses the Context-aware Distributed Sensor Network (DSN) for dementia patients. Incontinence in Patients with Dementia (PWD) due to a decline in their physical and mental abilities. Those PWD may lie in soiled diaper for prolonged periods if timely diaper change is not in place. Current manual care practices may not be able to immediately detect soiled diaper, although costly and labour intensive scheduled checks are performed. Delays in diaper change can cause serious social and medical issues. It uses different sensors like wet sensor, pressure sensor, and accelerometer to design DSN's which help in managing the PWD. Inputs from these sensors are also used to conserve battery power, either by reducing sampling rate of sensor or by switching off the sensor itself.

The work given in [12] presents combinatorial auctions for resource allocation in a distributed sensor network. This work discusses a solution to the problems posed by sensor resource allocation in an adaptive, distributed radar array. They have formulated a variant of the classic resource allocation problem, called the setting-based resource allocation problem, which reflects the challenges posed in domains in which sensors have multiple settings, each of which could be useful to multiple tasks. Further, they have implemented a solution to this problem that takes advantage of the locality of resources and tasks that is common to such domains. This solution involves translating tasks and possible resource configurations into bids that can be solved by a modified combinatorial auction, thus allowing us to make use of recent developments in the solution of such auctions. Also developed an information-theoretic procedure for accomplishing this translation, which models the effect various sensor settings would have on the network's output.

The work given in [13] describes the resource allocation for a distributed sensor network. In this work, they describe a project undertaken for the Office of Force Transformation (OFT) to investigate alternative resource allocation strategies for America's armed forces. In particular, OFT is interested in understanding how resource allocation strategies can be used in the context of distributed, network-centric units. To address this problem they have developed a simulation tool using agent-based modelling to explore the emergent properties of a distributed sensor network. To focuses on the task of using distributed sensors with varying characteristics and capabilities trying to detect and track the movement of enemy units in an urban environment. The goal of the project is to identify the impact of different resource allocation strategies on the performance of the sensor network.

The work given in [14] presents a resource allocation and congestion control in distributed sensor networks–a network calculus approach. The establishment of the overall objectives of a distributed sensor network is a dynamic task so that it may sufficiently well 'track' its environment. Both resource allocations to each input data flow and congestion control at each decision node of such a network must be performed in an integrated framework such that they are sensitive to this dynamically established overall objectives. In this paper, the effectiveness of a 'per-flow' virtual queuing framework that decouples the input data flows to each decision node is demonstrated. Under this framework, the buffer set point level of a decision node is established via the control of set point levels of individual virtual buffers assigned to each source node. Network calculus notions are utilized to model the end-to-end flow and design a simple yet effective feedback control law for each input data flow. The control strategy, while enabling satisfactory tracking of a dynamically allocated buffer queue set point, is also robust against the time varying nature of network delays and buffer depletion rate. Some of the works given in [15, 16, 17, 18]

DSN's always work under constrained resources. The resources can be any type like computational power, network bandwidth, battery power, memory or sensor node itself. Main motto in resource allocation is to increase the overall network life time of DSN's. Data caching, data aggression, context summarization, mobile agent based data collection are few important technique to improve network life time of DSN's by conserving resources.

4. PROPOSED WORK

Distributed Sensor Networks (DSN's) are mainly used for monitoring, tracking and remote controlling the system. We propose new context aware resource allocation of DSN's in retail chain management. We have taken Indian Retail industry as Case. DSN's are used in our proposed system to track the products in real time. Field Survey is done for major retail companies like More, Food World, Big Bazaar, Reliance, and Spencer. In the survey, we have found no real time information management system is available in market. And there is acute need for this kind of systems. We have received positive feedback from most of the store manager for our proposed system. Considering current retail industry, this section discusses about proposed system model, and functioning schema (proposed algorithm).

4.1 System Model.

In our proposed system, DSN's are deployed to obtain the real time information about availability of products in retail shop. We have developed model by considering only the weight sensors. But depending on type of the products to be tracked, DSN's may be designed to have one or more different types of sensors.

Figure2 shows the design of each rack. Each tray contains sensor which will keep track of how many items are available. This information is send to the cluster head (CH) in regular interval. The cluster head (CH) will also collect similar data from other racks. Data is aggregated in each cluster head before it is forwarded to another cluster head (CH).

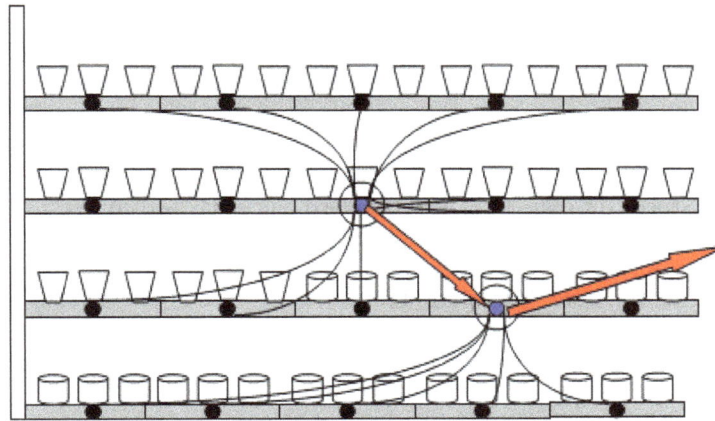

Figure2. Design of each rack.

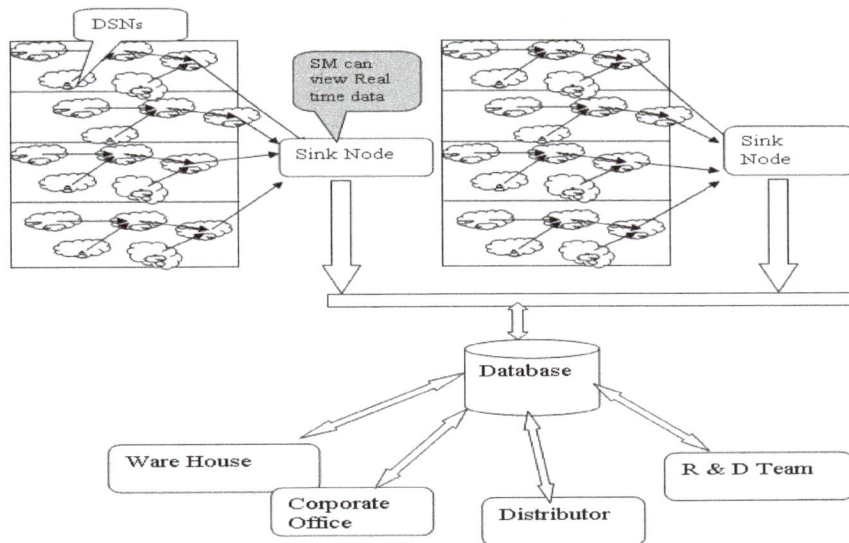

Figure3. Detail view of proposed DSN's system model for Retail Chain Management.

The Figure3 will explain proposed model. Data from each sensor in rack is forwarded to its cluster head. Each cluster head will aggregate the information sent by its sensor group and forward to its parent cluster. Forwarded information from all cluster head can be stored into a single system in each outlet shop. If there are many retail shops, then similar data from all retail shop can be stored in single centralized database. The store manager, corporate manager, distributor or R & D team can view this data for further analysis. Proposed system has following advantages compare to existing model 1) Store manager (SM) will be less depends on Floor Manager (FM) for any kind of updates 2) Product replacement within outlet for any floor can be done efficiently 3) No need of routine check up for each floor or rack 4) Labor cost can be reduced dramatically as fewer members are required for maintenance 5) Ware house manager can view the product availability at any time in each outlet. By this sold products can be replaced to each outlet depend on emergency. This process will to take at least one day in present system 6)Corporate head office can keep track of each and every outlet shop 7) Product Distributors will also get enough time to prepare for supply 8) R &D team can analyze the data obtained by this network for understanding consumer behavior, and 9) Reduces manual or paper work.

In our proposed system i.e. retail shop; we have derived many contexts in distributed sensor network environment. These contexts are used for efficient resource allocation. Main contexts are 1) Seasonal Context 2) Time Context and 3) Hybrid context.

In present working model of retail industry, Floor Manager (FM) has to do regular routine check up for availability of products in each rack. Figure5 discuss the advantage of DSN's over present system. The algorithm discussed below will explain how the burden on Store Manager(SM) is reduced by deriving meaningful data from context awareness of the system. Algorithm also explains about how to allocate resources depending on the context.

4.2 Steps for each Sensor Node in Proposed Work

Begin
1. Each sensor node is pre-configured to fixed weight and context. (Either Seasonal or Everyday context base).
2. After initial setup, each sensor nodes will sample data in a regular interval.
3. Depending on the context and available information, each node will derive an emergency value which signifies the attention to be given to that particular tray.
4. Derived value is forwarded to the Cluster Head (CH), which summarize further with other sensor nodes.
5. Sensor nodes can modify the event sensing rate or sleep time depending on available information, to reduce the power consumption.
6. Catering the service to every floor by Store manager (SM) solely depends on emergency value forwarded by DSN.
7. Steps 2 to 7 continue till the End of the Day (EOD).

End

4.3 Steps for each Cluster Node in proposed work

Cluster head forward the summarized emergency level to parent cluster head. Each Cluster Head will maintain data table related emergency level of its group. Update to table is done by each sensor node independently.

Begin
1. Initialize the cluster head to receiving mode.
2. Store the data in vector table.
3. Summarize the emergency value for Product information.
4. Forward the data to parent Cluster head in case of higher emergency.
5. Wait for next update from Sensor nodes in the cluster.
6. Repeat steps from 2 to 6.

End

4.4 Functioning Schema: Algorithm for each Sensor Node

We have used following data types and data structure for the algorithm. The algorithm has been developed using Castalia an OMNeT++ derivative. We have defined three type of context like time, seasonal, hybrid context. Three different scale and emergency level has been used. Structure *clusterRecord* is defined to store productId, rackId, emergency level to summarize the emergency information. *struct nodeRecord* store the static data for initializing the each node. This contains the table of information related to context, productID, rackID for each sensor.

Class *Connectivity Map* is designed according to Castalia requirement. This class is derived from VirtualApplication of Castalia publicly. VirtualApplication is generic class designed for development of application in Castalia. We also defined virtual functions like startup(), finishSpecific() and Application packet to exchange the data between the nodes. Following steps explain Implementation of Algorithm (using Castalia).

Step1. Once the basic initialization is done by Castalia, startup() is called for each node
Startup().
```
{
    Initialize variables specific to each node
        if(cluster node == true)
                Clear the cluster info.
        if(!Cluster Node)                                // Configure each normal node
                Configure_each_node()
        if ( nodeType = TIME_CONTEXT_SENSOR )
                setTimer(TIME_CONTEXT_TIMER, startTxTime);
        elseif(nodeType == HYBRID_CONTEXT_SENSOR )
                setTimer(HYBRID_CONTEXT_TIMER,txInterval_total*3);
                sendNodeToSleepState();
        else if ( nodeType == SEASONAL_CONTEXT_SENSOR )
                setTimer(SEASONAL_CONTEXT_TIMER, txInterval_total*9);
                sendNodeToSleepState();
}
configure_each_node()
{
        for each node i=1 to n
                intialise context type, productid, rackid, emergencyLevel.
}
```

Step2. After setting timeout in startup(), we wait for timeout. Check for sensor reading. If it is emergency send the packet to sink/cluster node.

```
timerOutCallback()
{
If( timeContextTimer) {
        emgLavel = getSensorValue();
        if(emgLevel == EMERGENCY_HIGH or EMERGENCY_MEDIUM )
                toNetworkLayer(packet2Net, SINK_NETWORK_ADDRESS);
        setTimer(TIME_CONTEXT_TIMER, txInterval_total);
}

If( hybridContextTimer)
{
        emgLavel = getSensorValue();
        if(emgLevel == EMERGENCY_HIGH or EMERGENCY_MEDIUM )
         toNetworkLayer(packet2Net, SINK_NETWORK_ADDRESS);
        setTimer(HYBRID_CONTEXT_TIMER, txInterval_total);
}

If (seasonalContextTimer)
{
        emgLavel = getSensorValue();
```

```
            if(emgLevel == EMERGENCY_HIGH or EMERGENCY_MEDIUM )
                    toNetworkLayer(packet2Net, SINK_NETWORK_ADDRESS);
            setTimer(SEASONAL_CONTEXT_TIMER, txInterval_total);
}
```

Step3. Cluster node will update the table, which should be forwarded to next cluster

```
fromNetworkLayer( )
{
     If (cluster Node)
          Update Emergency Level of cluster and forward to next cluster or sink
        If (sink node)
            If (emergencyLevel is high)
                Update to Store Manger and reduce Emergency Level of a node by
supplying products
}
```

Step4. While sending packet to network Layer, fill the product id, rackid, floor id and emergency level.

```
toNetworkLayer()
{
        Create ContextReportingPacket; Fill the productID, rackID, emgLevel, FloorID in
ContextReportingPacket, Send to Cluster Head
}
```

Step5. Collect the output of each module for further process using finishSpecific().

5. SIMULATION

We conducted simulation of the proposed scheme by using Castalia simulator (OMNeT++). The proposed model has been simulated in various distributed sensor network scenarios. For experimental purpose, we have conducted the simulation for 5 to 20 numbers of Nodes. We have considered M=1 to 5 Cluster and in each cluster we have considered N = 5 to 10 nodes. This section presents the simulation model, simulation procedure, results and discussions.

5.1. Simulation Model.

As discussed earlier, Castalia implements different modules of sensor node. Our simulations are conducted by developing separate application with required modification in sensor manager, radio, and power management module. In Castalia, node 0 is considered as SINK node and it also support bypass of routing and MAC layer to ease the testing in application layer. Since Context awareness is implemented in Application Layer, our application will bypass the routing and MAC layer and sent all packets to Node 0. The simulation is conducted with different number of sensor nodes. For simulation purpose, we have considered 5 to 10 nodes in each cluster. The simulation parameters are considered in this work ad shown in table1.

SINK Node	Node 0
Simulation Area	30 X 30 meters
Simulation Time	7000 seconds
Energy required for transmitting single packet [Tx mode Power]	46.2 mWatt
Energy required for receiving & processing single packet [Rx mode Power]	62 mWatt
Initial Power [Power of two AA batteries]	18720 Joules
Baseline Node Power [Energy utilized by node, even if node is not transmitting or receiving]	6 mWatt
Date Rate	250 kbps
Modulation Type	PSK
Bandwidth	20 MHz

Table1. Simulation Parameters.

5.2. Simulation Procedure.

To illustrate the results of simulations, we have setup the configuration files of OMNeT++ i.e. .ini and .ned file. In all the scenarios, Node 0 is considered as SINK node with one hop reaching. This is because Castalia simulation tool except all nodes to be one hop count reach from the sending node. We have conducted simulation of the proposed scheme by using Castalia simulator (OMNeT++). For experimental purpose, we have conducted the simulation for 5 to 20 numbers of Nodes. We have considered M=1 to 5 Cluster. And in each cluster we have considered N = 5 to 20 nodes. Node 0 is considered as SINK node as in Castalia. This section presents the simulation model, simulation procedure, results and discussions. To illustrate the results of simulations, we have setup the configuration files of OMNeT++ i.e. .ini and .ned file. Following are the major default values used in our simulation. Consider the following steps involved in the simulation.

Begin
1. Initialize .ini and .ned files with parameters along with the required default value.
2. Configure each sensor node according to application requirement. Each node has to be configured for Context Type, Product ID, Rack ID and default emergency Level.
3. Set Context Timer for each node, compile and run the simulation with Castalia script. In our simulation we have set 200s, 600s, 1800s for time context, Hybrid context and seasonal context sensor.
4. After timeout, each node sends the data to SINK node. Required data is each layer is collected by collectOutput() in finishSpecific().
5. Use CastaliaResult utility to analyze the collected information.
6. Compute the performance parameters.
7. Generate the Graphs according to results obtained.

End

5.3. Performance Parameters.

We have used the following parameters to measure the performance of the proposed scheme.

1. **Number of Packets Transmitted (Tx):** The packet transmission is defined as the total number packets transmitted by each nodes in different contexts.
2. **Energy Utilization:** It measures the energy utilization of each node in different context as measured in joules.
3. **Time taken to reduce the Emergency Level:** It measures the time required to reduce the emergency level from high to low by nodes in distributed sensor network environment.
4. **Battery Energy Consumption:** It is the battery energy consumption of all nodes in sending, receiving and forward operations. A small amount of energy also lost during state transition and listening mode.
5. **Throughput:** It is the ratio of number of packets sent to the number of packets received by all nodes. Throughput is c calculated for different type of sensor.
6. **Resource Allocation:** It measures the percentage of resources allocated successfully to all the nodes in different contexts. It is measured in percentage.

5.4. Results and Discussions

5.4.1. Number of Packets Transmitted (TX) Vs Number of Nodes.

Figure4 shows that total number of packets sent by all nodes. Number of packets sent for time context is much higher compare to seasonal context and hybrid context. Figure5 shows the graph about total number of packets which are not received due to sleep mode of nodes. Since more packets are sent in time context, it shows maximum packets. Processing each packet will consume 62mw of energy. To avoid this we are sending each node to sleep mode till its timer expires.

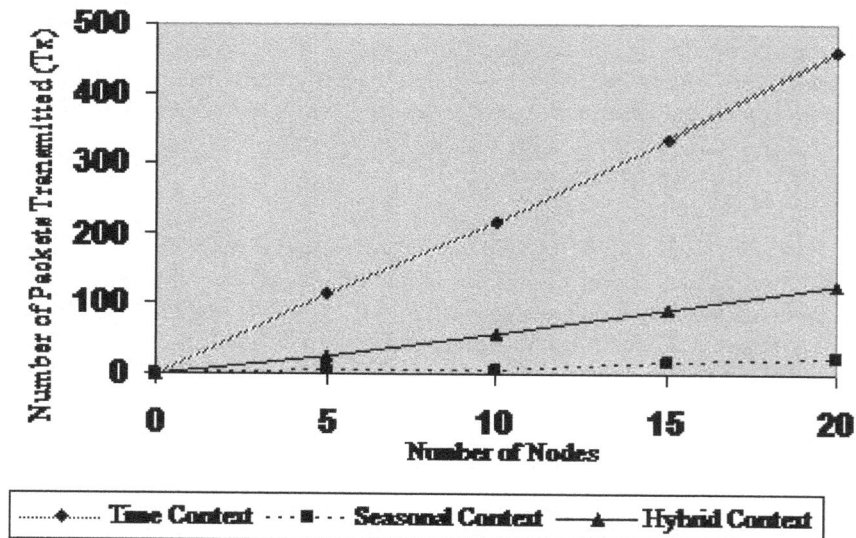

Figure4. Number of Packets transmitted (Tx) Vs Number of Nodes.

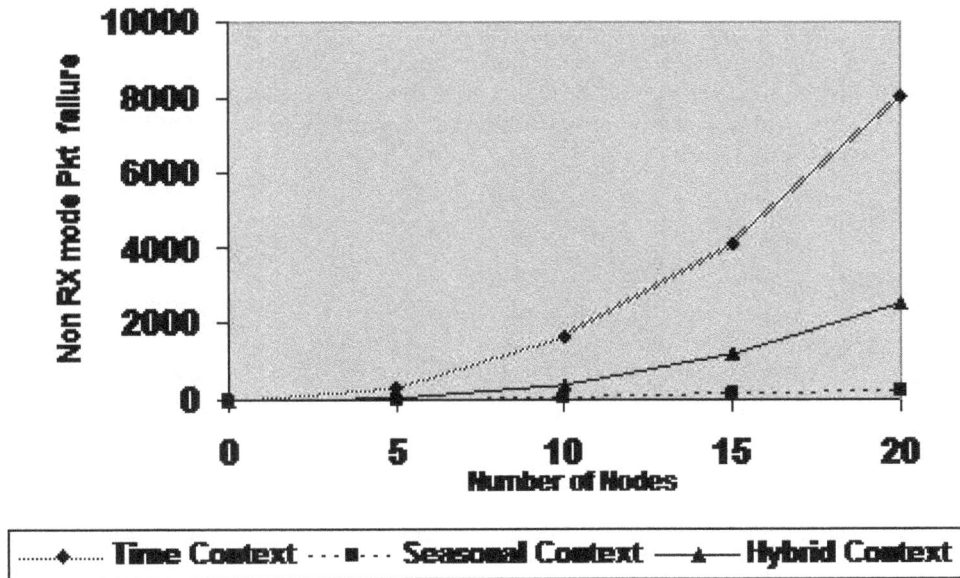

Figure5. Non RX mode Pkt failure Vs Number of Nodes.

5.4.2. Energy Utilization Vs Number of Nodes.

Figure6 shows the graph about average amount of energy spent in each context. As in graph the energy spent in time context is always much high compare to other context. Since overall packets sent in seasonal and hybrid context is less, energy spent is also much lesser compare to time context.

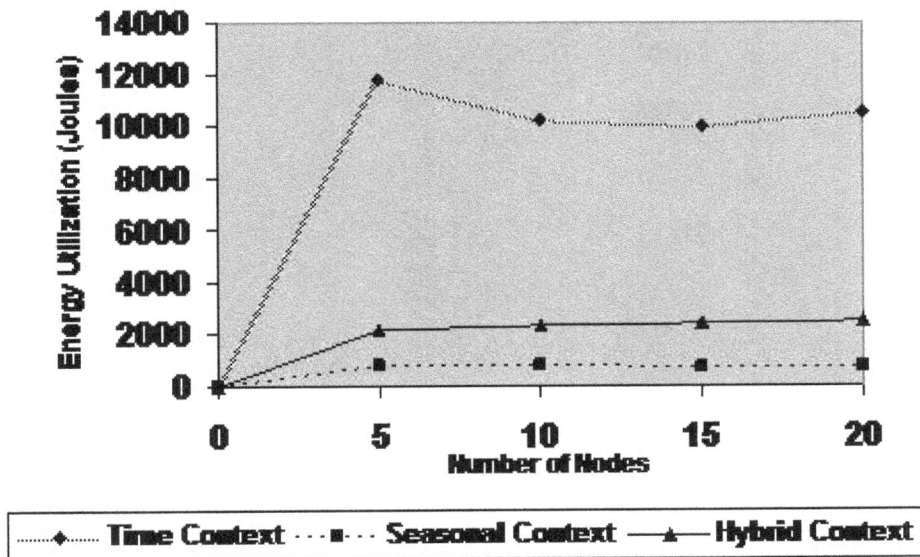

Figure6. Average energy Utilization by all nodes Vs Number of Nodes.

5.4.3. Energy Consumption.

Figure7 shows the battery consumption in different context. In time context sensor nodes, power consumption is almost linear and it is much higher compare other context nodes. The power consumed by hybrid context sensors is less than a 25% that of time context sensor node. Energy consumption by Seasonal sensor node is also much lower than time and hybrid sensor node. (14% of time context sensor node).

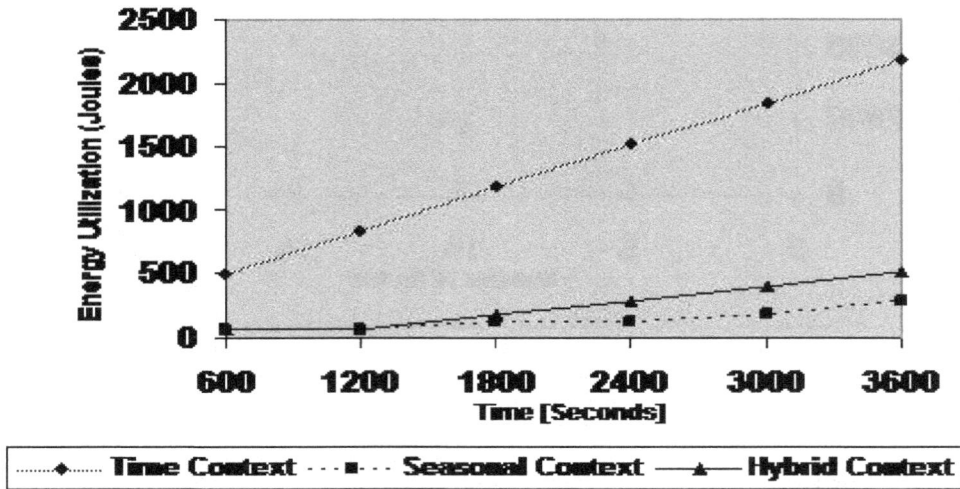

Figure7. Energy Utilization Vs Time.

5.4.4. Time taken to reduce Emergency Level.

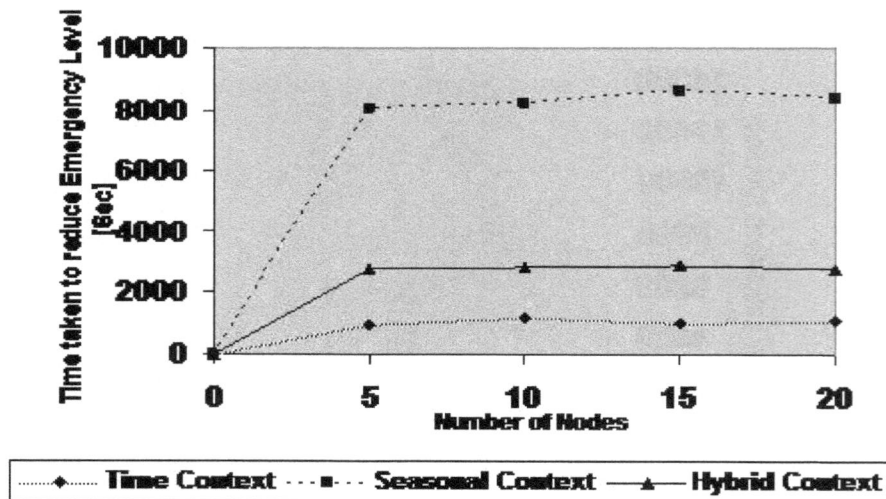

Figure8. Time taken to reduce emergency level Vs Number of Nodes.

Figure8 shows the time taken to reduce the energy level in each context. Generally products related to time context (Products which are daily usage) will be sold quickly. So these products has to be deployed much earlier compare to seasonal and hybrid context.

5.4.5. Throughput.

Figure9 shows the throughput of nodes in different context. Throughput is calculated as ratio of number of packet received to the number of packet sent by all nodes. Sensor nodes in seasonal context outperform the time and seasonal context. This is because number of packets sent in seasonal is less compare to seasonal and hybrid context.

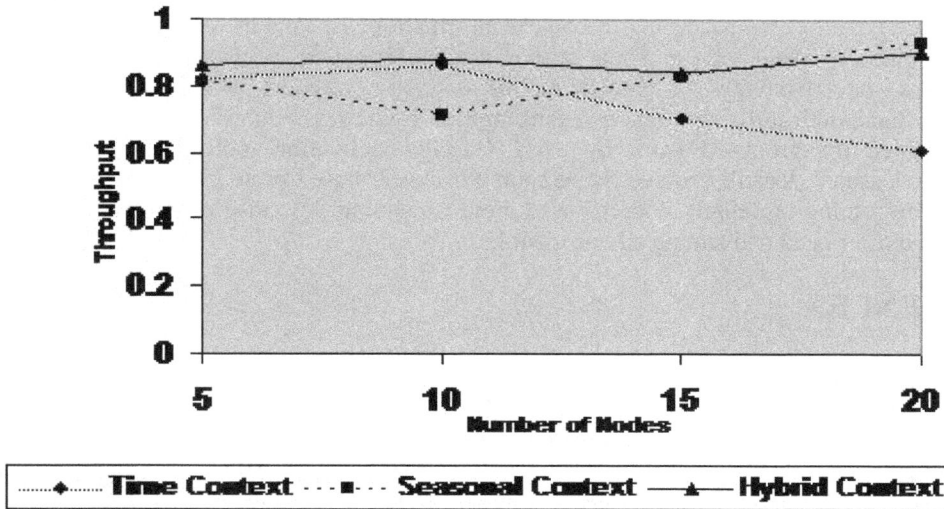

Figure9. Throughput Vs Number of Nodes.

5.4.6. Percentage of Resource Allocation.

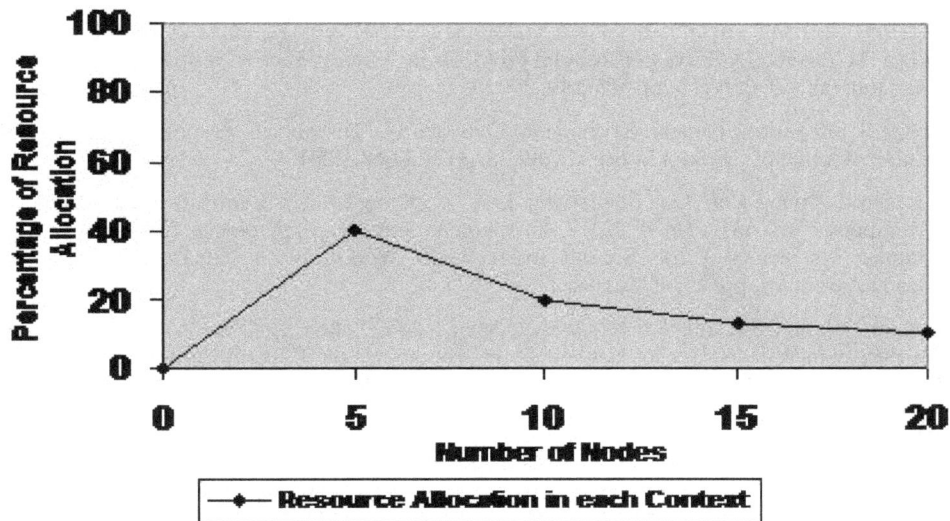

Figure10. Resource allocation Vs Number of nodes.

Figure10 shows the percentage of resource allocation in any context. Here numbers of active nodes are considered as resource. At any point of time, only cluster node and transmitting nodes are activated in any cluster group. The remaining nodes are pushed to sleep state. Because scheduled activation, energy consumption considerable reduced.

6. CONCLUSION

Context aware systems are designed using human perspective, domain specific information. We have analyzed drawbacks of Indian Retail Industry and then we have proposed new system using DSN's. In our design, sensors are deployed in each rack which provides the real time information of products available in that rack. This information can be used by warehouse manager, Store manager for day to day activity in super market or hyper market. R&D team can also use this information to understand consumer behaviour. Sensor nodes and other network resource are also allocated according to context information. To save the energy further, nodes are programmed to sleep state as soon as they send information to cluster node. Simulation results show the advantages of embedding the context awareness in system for resource allocation. Bandwidth utilization, energy consumption is greatly reduced in each context. We have simulated the proposed work by using Castalia simulation tool which is based on OMNeT++. Castalia doesn't provide the support for cluster based protocol and also it support only one hop routing in default mode. Due to these constraints, we have considered sink node (node 0) as cluster head and simulated our module.

REFERENCES

[1] S.S., Iyengar, Ankit. Tendon, and R.R., Brooks, "A overview of Distributed Sensors Network".

Available From:

http://books.google.com/books/about/Distributed_sensor_networks.html?id=Nff5

[2] Shivakumar Sastry and S.S., Iyengar, "Taxonomy of Distributed Sensors Network".

Available From:

http://books.google.com/books/about/Distributed_sensor_networks.html?id=Nff5

[3] Jamal. N, Al. Karaki, Ahmad. Kamal, "Routing techniques in wireless sensor Networks: A survey, Journal of IEEE Wireless Communications, vol. 11, No.6, pp 6-28, 2004.

[4] K. Akkaya, M.Younis, "A Survey of Routing Protocols in Wireless Sensor Networks". In Elsevier Ad Hoc Network Journal, vol. 3, No. 3, pp 325-349, 2005.

[5] Hairon. Qi, S. Sitharama., Iyengar, Krishnendu. Chakrabarty, "Distributed sensor networks - review of recent research". Journal of Franklin institute, vol.338, pp 665-668, 2001.

[6] Faraz. Rasheed, Young-Koo. Lee, Sungyoung. Lee, "Applying Context Summarization Techniques in Pervasive Computing Systems". The 4th IEEE Workshop on Software Technologies for Future Embedded and Ubiquitous Systems, and the Second International Workshop on Collaborative Computing, Integration, and Assurance, pp. 107-112, 2006.

[7] Suan. Khai., Chong, Shonali. Krishnaswamy, Seng. Wai., Loke, "A Context-Aware Approach to Conserving Energy in Wireless Sensor Networks". In proceedings on 3rd IEEE Conference on Pervasive Computing and Communications, pp 401 – 405, 2005.

[8] Available From;

http://www.omnetpp.org/doc/omnetpp41/manual/usman.html#sec101

[9] Chong, S. K., Krishnaswamy, S., Loke, S. W., "A context-aware approach to conserving energy in wireless sensor networks". In proceedings of 3rd International Conference on Pervasive Computing and Communications Workshops, pp. 401-405, 2005.

[10] Anand. Agarawala, Saul. Greenberg, Geoffrey. Ho, "The context-aware pill bottle and medication monitor department of computer science". In proceedings of video proceedings and proceedings supplement of the UBICOMP 2004.

[11] Aung. Aung., Phyo.., Wai, Foo. Siang., Fook, Maniyeri. Jayachandran, Jit. Biswas, Jer-En. Lee, Philip. Yap, "Implementation of Context-Aware Distributed Sensor Network System for Managing Incontinence among Patients with Dementia". In proceedings of International conference on Body Sensor Networks, pp 102-105, 2010.

[12] J. Ostwald, V. Lesser, S. Abdallah. "Combinatorial Auctions for Resource Allocation in a Distributed Sensor Network". In proceedings of the 26th IEEE International conferences on Real-Time Systems Symposium, (RTSS'05), IEEE Computer Society Press, pp 274 – 279, 2005.

[13] Martin. C., Martin, Iavor. Trifonov, Eric. Bonabeau, "Resource allocation for a distributed sensor network", 2005.

[14] J. Zhang, K. Premaratne , Peter H. Bauer, "Resource allocation and congestion control in distributed sensor networks–A network calculus approach". In Proceedings of 15th International Symposium on Mathematical Theory of Networks and Systems, 2002.

[15] Zhikui. Chen, Zhe. Wei, "Intelligent Home-Hospital System Based on Context-Aware Technology". In proceedings of international conference on Industrial and Information Systems, pp 23- 26, 2009.

[16] Hristova, A. Bernardos, A.M. Casar, J.R., "Context-aware services for ambient assisted living: A case-study". In proceedings of Applied Sciences on Biomedical and Communication Technologies, pp 1-5, 2008.

[17] HK Pung, T Gu, Wenwei Xue, PP Palmes, J Zhu, WL Ng, CW Tang and NH Chung, "Context-Aware Middleware for Pervasive Elderly Homecare". IEEE Journal on Selected Areas in Communications, vol. 27, no.4, pp 510 -524, 2009.

[18] Tomoya. Kawakami, Bich. Lam., Ngoc.., Ly, Susumu. Takeuchi, Yuuichi. Teranishi, Kaname. Harumoto, Shojiro. Nishio. "Distributed Sensor Information Management Architecture Based on Semantic Analysis of Sensing Data". In Proceedings of SAINT, pp 353 -356, 2008.

12

GROUP BASED ALGORITHM TO MANAGE ACCESS TECHNIQUE IN THE VEHICULAR NETWORKING TO REDUCE PREAMBLE ID COLLISION AND IMPROVE RACH ALLOCATION IN ITS

[1]Ramprasad Subramanian, [2]Shouman Barua, [3]Sinh Cong Lam, [4]Pantha Ghosal, [5]Kumbesan Sandrasegaran

[1,2,3,4,5]Centre for Real-time Information Networks, School of Computing and Communications, Faculty of Engineering and Information Technology, University of Technology Sydney, Sydney, Australia

ABSTRACT

Intelligent transportation system (ITS) is an application which provides intelligence to the transportation and traffic management systems. Although the word ITS applies to all systems in the transportation but as per the European union directive it is the application of Information and communication technology in the field of transportation is defined as ITS. The communication technology has evolved greatly today from 2G/3G to long term evolution (LTE). In this paper we focus on the LTE and its application in the ITS. Since LTE offers excellent QoS, wide area coverage and high availability it is a preferred choice for vehicle to infrastructure (V2I) service. At the same time the LTE customer base is increasing day by day which results in congestion and accessing the network to send or request resources becomes difficult. In this paper we have proposed a group based node selection algorithm to reduce the preamble ID collision otherwise this uncoordinated preamble ID transmission by vehicle node (VN) will eventually clog the network and there will be a massive congestion and re-transmissions attempts by VNs to obtain the random access channel (RACH).

KEYWORDS

Intelligent transportation system (ITS), Long term evolution (LTE), Mobile ad hoc network (MANET), Vehicle ad hoc network (VANET), Vehicle to infrastructure (V2I), Vehicle to vehicle (V2V), Random access channel (RACH).

1. INTRODUCTION

Intelligent transportation system (ITS) refers to the application of modern telecommunication technology in the control of the transportation system. The time spent by the people in the cars and in other transportation has increased [1] and many people prefer driving themselves in the long weekend rather than taking up the public transportation. So the modern ITS should encompass of automated highways, automated toll collection system, vehicle tracking system, intelligent transportation and logistics, in-vehicle GPS and mapping systems, automated enforcement of traffic lights and speed laws, smart control devices[2]. But the key to make the transportation systems intelligent is made possible with the application of telecommunication technology in the transportation domain. The word transportation systems became intelligent transportation system with the application of telecommunication technology. The long term goal of the ITS is to make the transportation system more and more autonomous with the help of the telecommunication technology[2]. This long term goal also provides a huge challenge to the telecommunication technology to further grow in the areas of robust technology, high data rates, with adequate coverage etc. But this long term goal should be ably supported by lot of short term goals which can be realised with the current advancement in the telecommunications. These short term goals include making the roads more and more efficient and safer to travel by fulfilling the growth in the following areas:

a. Blind spot detection.
b. Collision avoidance.
c. Intelligent navigation using traffic light updates.
d. Intelligent traffic control using real time traffic information.

The medium term goals and opportunities leads to autonomous driving:

a. Provision of the telecommunication infrastructure support for the autonomous driving.
b. The telecommunication is the fulcrum of the autonomous driving and without that, achieving the autonomous driving in a large scale is not feasible.
c. Traffic control and navigation in a large dynamic environment is not feasible without the communication technology support.

The long term goals include:

a. In car office as indicated.
b. In car entertainment and many more.

Telecommunication is the key to make ITS happen and ITS provides tremendous opportunity for the growth in the telecom sector. There is a sort of serendipitical relationship exists between the telecommunication and ITS. For example in developing countries such as India where lot of people travel in their cars to reach the office because but the problem is heavy traffic congestion and as a result of this the quality man hours is wasted in the traffic and the productivity is affected. So in order to overcome this problem telecommunication can be effectively used to control the traffic and to provide all the latest infrastructure of the office environment inside the car as long term goal of the ITS. This will enable the people to start the work immediately once when they get into the car and the quality man hours can be fully utilized.

In the interaction between the ITS and telecommunications, the later should come up with the customized solution to meet the ITS requirement. At the same time the information delivered by the telecommunication systems to the ITS system must be handled properly and with some strategy. Otherwise, even with the information there will not be any improvement in the system.

2. COMMUNICATION NETWORK DESIGN FOR ITS

Several network designs and several protocols have been proposed by various researchers in the telecommunications over past few years to enable ITS and its application happen. But still there is no solution for all the needs. So an important question may arise at this juncture why we need so many forms of communication systems for the ITS[3]. The answer is simple to this question. The applications of ITS is not just in one area to provide one full proof system to cater the need[2],[3]. The applications of ITS are numerous so based on the intended applications the telecommunication systems can be remodelled. Likewise for vehicular networking there exists two methods of communication setup and they are vehicle to vehicle (V2V) and vehicle to infrastructure (V2I). For the communication links between V2V numerous algorithms have been developed and specified by various researchers. Apart from this IEEE has standardised the

VANET with IEEE 802.11p1 standards which communicates between the vehicles[4],[5]. VANET is particularly designed for the short range communications between the vehicles. Since there is no infrastructure support for the VANET the communication range cannot be extended beyond certain limits. This technology offers a tremendous networking capacity between the vehicles but when it comes to long range communication needs then instead of VANET, LTE would be an appropriate choice. The focus of this paper is LTE in V2I.

The V2I architecture is the communication link between the vehicle and infrastructure. In this the vehicle can communicate with the content server located with the service provider's network to fetch the required information. For example if a person from country A is going to country B and happens to drive a car. The geography of the new place will be alien to him. So in order to reach the destination properly he or she can request the route information to the content server from the vehicle and the trip planner can guide him properly to reach the planned destination. This cannot be achieved by using V2V architecture and instead V2I will be useful. Apart from this if an accident happens in a bridge the message of the accident has to be informed to the intended users of the bridge and propose an alternative route to them in order to control the traffic jams because of the accident so that the commuters can take the alternative route to reach the destination. This type of network controlled operations can be performed using V2I architecture and the same is not possible with V2V architecture.

V2I architecture can be effectively used by the emergency service providers for example in a situation where a person is travelling in a motorway and if somebody is experiencing an emergency and needs immediate attention or help, then the person can propagate the appropriate request message to the emergency handling centre. Not only in the emergency condition V2I is also very useful in lot of other circumstances like in a situation where a guidance is required from the expert, requesting information from the ITS service provider data base etc. There is some drawback in this V2I architecture apart from the advantages specified previously. In this architecture each time a person A propagates the information to person B even if person B is geographically located close to person A the information has to take a long route of going through the central server from the vehicle. So this will result in some delay for the information to reach the destination.

2.1. LTE for V2I architecture in ITS

LTE is a evolution of 3G UMTS. The main improvement of the LTE from its predecessors is the removal of base station controller (BSC) or radio network controller (RNC). The intelligence of the base stations in 2G and 3G is limited and they are mainly controlled by BSC and RNC. These controllers play a major role in radio resource management, call assignment procedure and control of base station nodes. Apart from this the controllers are controlled by circuit switched network (CS core). The 2G network has very minimum data capacity. The 3G system which got evolved from 2G offered a better data capacity compared to 2G. But the LTE/LTE-A which got evolved from 3G offers a excellent data rate capacity of 1Gbits/s in peak download and 500 Mbits/s in upload.

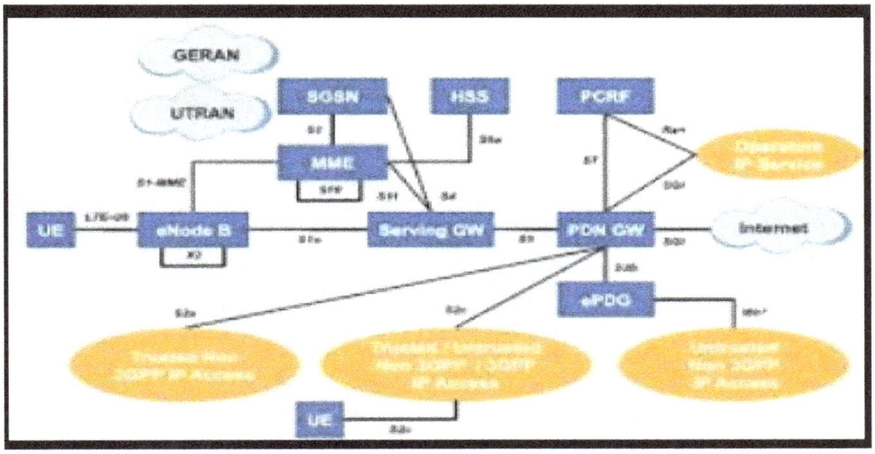

Figure 1. LTE architecture (Alcatel-lucent)

The network architecture of LTE doesn't have any similarities with 2G or 3G. In LTE there is no concept of BSC or RNC. The nodeB's (NB) from 3G got evolved into eNodeB (eNB). The eNB are connected to mobility management entity (MME). MME is an evolved packet core (EPC) element. The MME resides in the EPC control plane and manages mobility management activities like session states, authentication, paging, mobility with 2G and 3G nodes, roaming, and other bearer management functions. The EPC differs from the CS core and packet switched core (PS core) in many aspects. The EPC routes the packets through internet protocols (IP). It supports both IPv4 and IPv6. The EPC always maintains the IP connection between the mobile and the outside world by setting up a basic IP connection. This feature of LTE differs with 2G and 3G. The connections are made when it is requested and after the session is closed the connection to the outside world is disconnected.

The EPC behaves as a data pipe between the external world and to the mobile. It just transports the information to and from the external world to the mobile and vice versa. This operation of the EPC is similar to that of the normal internet connection. EPC does not care about the content of the packets. It just transmits all the information inside the pipe. In EPC the voice application is not the part of the system. It is handled separately by IP multimedia system (IMS). This operation of EPC varies with the traditional telecommunication networks in which voice forms an integral part of the network. The EPC simply transports the packets which contains voice packets similar to other data packets. The EPC has the mechanism to control and specify the data rate, error rate and delay to travel across the EPC. There is no timing requirement for the data packet to travel across the EPC in user plane but the specifications suggests that 10 milliseconds for the normal mobile and 50 milliseconds for the roaming mobile. The EPC should also support the handovers between the 2G and 3G systems.

The table below shows the different features and the associated network elements in LTE and UMTS and suggests the difference between them.

Table 1. UMTS and LTE network elements details

Feature	UMTS	LTE
Radio access network components	NodeB, RNC	eNB
RRC protocol states	CELL_DCH, CELL_FACH, CELLPCH, URA_PCH, RRC_IDLE	RRC_CONNECTED, RRC_IDLE
Handovers	Soft and hard handovers	Hard
Neighbour lists	Always required	Not required

The LTE/LTE-A is designed to handle 1Gbits/s in download and 500 Mbits/s in the upload. This tremendous capacity lured the ITS and its application to adopt this technology as the technology for backbone communication in V2I infrastructure service. The data rate capacity of LTE is attracting more and more people to migrate to LTE from other traditional technologies like 2G and 3G. As a result of this there is a huge increase in the customer base and in turn congestion in the network.

Many developed countries across the world are slowly introducing ITS and its applications in the traffic management. Apart from the government agencies the vehicle manufacturers like Toyota, BMW, GM etc are introducing lot of ITS features in the vehicles. Apart from supporting ITS the operators are introducing new features and they are supporting lot of machine to machine (M2M) services in order to increase the revenue to compensate for the increase in opex. As per the ETSI survey [6] around 50 billion machine to machine devices are expected to occupy the market in 2020. But this comes with the price of increased congestion in the random access channel (RACH). But in our paper we will restrict our discussion to the VNs. These VNs will be in the idle mode when they don't have anything to transmit and become to active mode when it is transmitting any information. To become active, the VNs have to request for RACH by sending a preamble ID. But as per the analysis as the number of VNs increase the collision percentage of the preamble ID also increases. So as a result the VNs will go for retransmissions and it will finally end up in clogging the network.

The data that has been presented in Figure 2 is the result of collection of RACH counters to analyse the RACH failures for past three months from an operator LTE network and in which more than a billion of RACH attempts where studied and from that RACH failure percentage has been calculated. The results below shows only the RACH failures in the network. From this below Figure 2 we can attribute that 30% of the RACH attempts in the networks is failing and only 70% of the RACH attempts are successful. Through this analysis we want to confirm that LTE network is already getting clogged up without much of the usages in the ITS applications or other M2M services as of today and the situation will get worse if we start supporting these service. At the same time a proper random access technique should be addressed to improve the situation as the current techniques has many shortcomings.

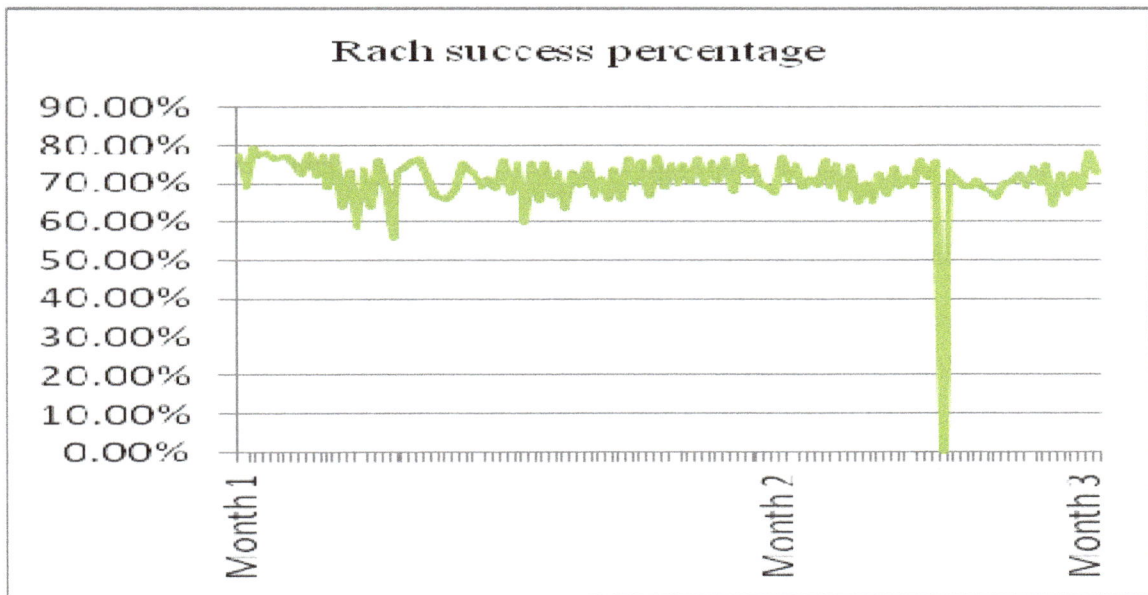

Figure 2. RACH KPI counter from live network

3. RELATED WORK

Many algorithms has been proposed earlier to overcome the RACH congestion. These algorithms can be classified as Delay focused, Non-hierarchical, success rate, non-hierarchical and energy based. [7] proposes cluster based approach to manage access control of massive kind. Jitter is considered as the main QoS criteria. In [8] proposes access management technique based on the events. This article proposes fast adaptive SALOHA scheme for this. [9] proposes collaborative access class barring approach. In this technique if M2M device service area covers more than one BSs, then access class barring scheme for the BSs will be modified based on the congestion of the BS. Radio access network (RAN) control method for synchronised M2M traffic is proposed in [10] because synchronized traffic is exerts more load in the network than asynchronous traffic. [11] proposed the random scheme based on the network congestion level access class barring scheme is implemented. Here M2M traffic is subdivided into five major classes and priorities are assigned accordingly. In [12] a new scheme has been proposed that does not have influence on H2H services. The M2M device remembers successful contention information to achieve contention free RACH. All this algorithms proposed are for static M2M devices. Even though vehicular nodes are classified among this M2M devices but they are highly dynamic nodes. Hence the proposed algorithms have limitations when it comes to the dynamic M2M devices. Hence a group based algorithm to manage access technique in vehicular networking to reduce preamble ID collision has been proposed.

4. PROPOSED ALGORITHM

As seen in the Figure 3 the vehicle devices will communicate with each other using IEEE 802.11 P1. Each device will try to behave as a group leader thinking that the LTE signal received by the VN is the strongest. The received signal strength of the network is constantly exchanged between the VNs using IEEE 802.11 p1 signal. During the exchange if one of the VN identifies that the received signal from the neighbour node is greater than that of the recipient node then the recipient node becomes a member of the group headed by the transmitter node. The node with highest signal strength (signal from LTE cell) becomes the group leader and other nodes who join the group becomes the group members. The groups created are amorphous in nature and the group leader can change swiftly due to the highly dynamic nature of the RF environment and also due to highly dynamic nature of the vehicles. If the leader is changed then the group collapses and new leader creates new group with his members. There is no limiting capacity to number of members who can join in a group. Since there is every chance that group leader will change soon the number of node limitations in the group is not required. The group leader monitors the weighted values of its members and if the weighted values falls below certain level the node will be released from the group by the group leader.

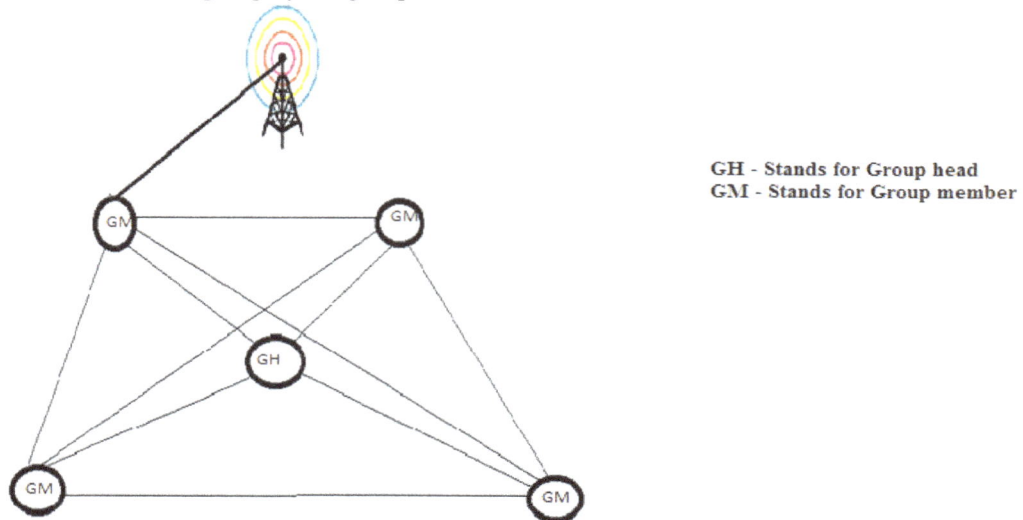

GH - Stands for Group head
GM - Stands for Group member

Figure 3. Schema of the proposed algorithm

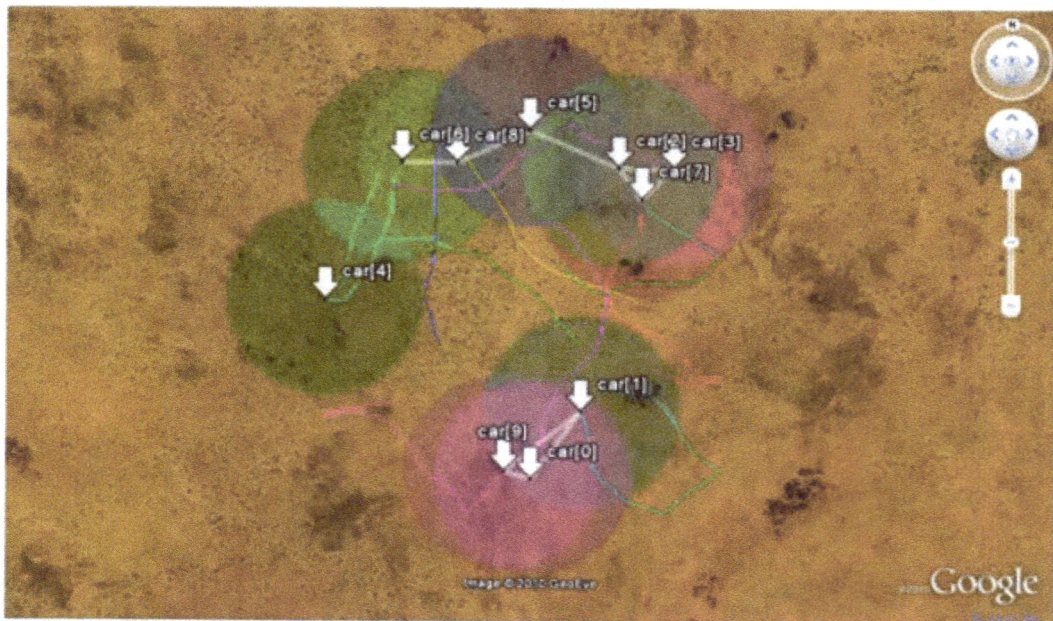

Figure 4. Simulation of the grouping proposed in the algorithm

Apart from the signal strength, the other factors such as direction of arrival (DoA), position and the velocity of the group members are calculated and based on this weighted values are allocated for each group members and the member will be arranged in the descending order and will be allocated with a time slot to transmit the same preamble ID used by the group head earlier. Once the group head successfully transmits its preamble ID without any collision then same preamble ID will be used by the group members to as reach the eNB.

In our algorithm we assume that both the group members and the group leader are in motion and the data we receive from the members are not homogeneous. In an environment in which the fading characteristics are rapidly changing more instantly, estimating the correlation matrix is computationally intensive to user other DoA methods such as ESPRIT, MUSIC etc. So that's why the non-statistical or direct data domain (D3) technique known as Matrix Pencil algorithm to estimate the DoA of the signal has been chosen. In our Algorithm we assume that all the VNs are transmitting at the constant power level. So based on the received signal strength and based on the DoA of the signal the position estimation can be derived. But this is just an approximate position estimation and some tradeoffs can be allowed in this calculation. Since the position estimation carries less weightage as compared to signal strength from LTE cell and DoA. The exact coordinates of the vehicles obtained in vehicle navigation systems from GPS will not be used to maintain the privacy of the individual. The group leader will poll its members at a regular intervals and the difference in the position between the first poll and second poll will be used to calculate the velocity of the member. Each VN will transmit its LTE signal strength information to other VN and each node will be receiving this information from other nodes. The vehicle nodes uses triangulation technique to find out its position and this will be relative to geographic north. This self position info calculated by the node will be used if that node become the leader and form its own group.

The algorithm is represented in the form of a flow chart in the figure below.

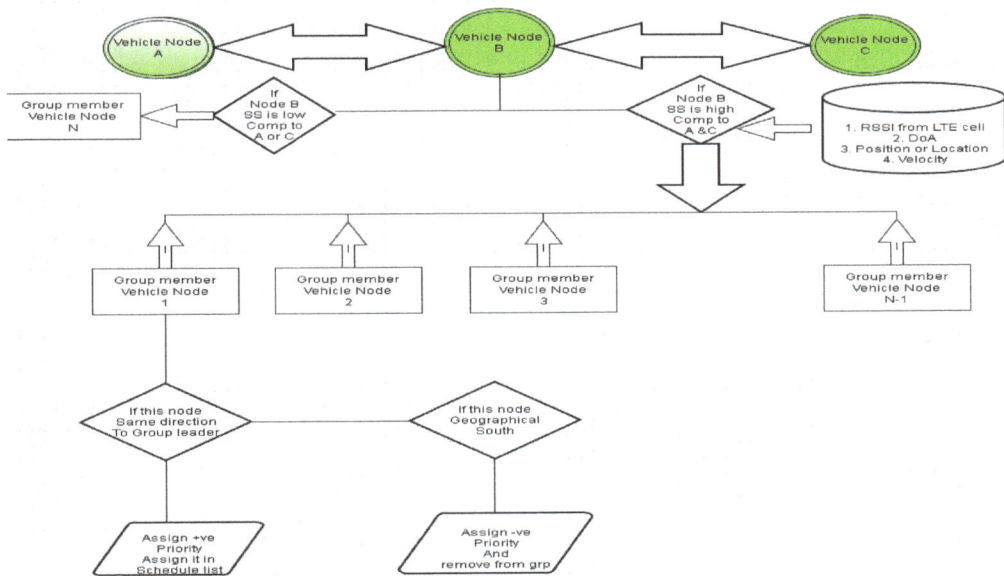

Figure 5. Flow chart of the proposed algorithm

4.1. Algorithm design

4.1.1. Calculation of directional of arrival using matrix pencil theorem

The algorithm like MLE, MUSIC and ESPRIT calculates the DoA based on the correlation matrix R. To construct a correlation matrix these algorithms consumes considerable amount of computational load because of the correlation and this makes it more complex to use the and especially in a environment which is highly dynamic. To estimate the correlation matrix we need at least K samples from the data x where $K > 2N$. The K samples of the data can obtained from the K snapshot from the target under the consideration. But with the prime assumption that all the K samples follow the same statistics i.e., the data is homogenous. In an environment in which the fading characteristics are rapidly changing and waiting for K samples of data and then calculating the matrix to estimate the DoA may not time consuming and computationally intensive. The proposed algorithm works in a highly dynamic environment and hence decided to choose matrix pencil theorem which works in the non-statistical way and computes the DoA with the data it receives unlike the other algorithms which waits for the K samples of data to estimate the DoA. Matrix Pencil was originally developed for the estimation of the poles of a system. However, it can be applied as well to DOA estimation. In the original Matrix Pencil the received data at time index n is given by

$$x_n = \sum_{m=1}^{M} A_m z_m^n + n_n \tag{1}$$

where $z_m = e^{j\,kd\,\cos\,\varphi_m \Delta t}$ represent the poles of the system, n_n represents the AWGN. The goal is to estimate z_m given x_n, $n = 0, \dots N - 1$.

To improve the position accuracy of the group members calculation we have assumed that VNs uses multiple antenna terminals. So in our case, the data is received at the group head terminals of N antenna elements from the group member, otherwise the formulation is exactly the same. Hence, the original Matrix Pencil algorithm is applicable to DoA estimation. The matrix Pencil algorithm which we have chosen here has many similarities to the other DoA estimation techniques such as ESPRIT, but without estimating a correlation matrix. We begin by defining two $(N - L) \times L$ matrices X_0 and X_1 as

$$\mathbf{X_0} = \begin{bmatrix} x_0 & x_1 & \cdots & x_{L-1} \\ x_1 & x_2 & \cdots & x_L \\ \vdots & \vdots & \ddots & \vdots \\ x_{N-L-1} & x_{N-L} & \cdots & x_{N-2} \end{bmatrix} , \quad \mathbf{X_1} = \begin{bmatrix} x_1 & x_2 & \cdots & x_L \\ x_2 & x_3 & \cdots & x_{L+1} \\ \vdots & \vdots & \ddots & \vdots \\ x_{N-L} & x_{N-L+1} & \cdots & x_{N-1} \end{bmatrix} . \tag{2}$$

where L is a pencil parameter that must satisfy

$$M \le L \le N - L \quad N \text{ even} \tag{3}$$

$$M \le L \le N - L + 1 \quad N \text{ odd}. \tag{4}$$

The basis of Matrix Pencil is that, based on the data model, we can write these matrices as

$$X_0 = Z_1 A Z_2, \qquad (5)$$

$$X_1 = Z_1 A \Phi Z_2, \qquad (6)$$

where Φ is the same as in ESPRIT, the diagonal matrix that we want to estimate. The four matrices are given by

$$Z_1 = \begin{bmatrix} 1 & 1 & \cdots & 1 \\ z_1 & z_2 & \cdots & z_M \\ \vdots & \vdots & \ddots & \vdots \\ z_1^{(N-L-1)} & z_2^{(N-L-1)} & \cdots & z_M^{(N-L-1)} \end{bmatrix}_{(N-L) \times M}$$

$$Z_2 = \begin{bmatrix} 1 & z_1 & \cdots & z_1^{L-1} \\ 1 & z_2 & \cdots & z_2^{L-1} \\ \vdots & \vdots & \ddots & \vdots \\ 1 & z_M & \cdots & z_M^{L-1} \end{bmatrix}_{M \times L}$$

$$\Phi = \begin{bmatrix} z_1 & 0 & \cdots & 0 \\ 0 & z_2 & \cdots & 0 \\ \vdots & \vdots & \ddots & \vdots \\ 0 & 0 & \cdots & z_M \end{bmatrix}_{M \times M}$$

$$A = \begin{bmatrix} \alpha_1 & 0 & \cdots & 0 \\ 0 & \alpha_2 & \cdots & 0 \\ \vdots & \vdots & \ddots & \vdots \\ 0 & 0 & \cdots & \alpha_M \end{bmatrix}_{M \times M}$$

Without noise, for the choice of pencil parameter L that satisfies the constrains in eqn. (4), the matrices X0 and X1 have rank M. Consider the matrix pencil $X_1 - \lambda X_0 = Z_1 A [\phi - \lambda I] Z_2$. For arbitrary λ, this matrix difference also has rank M. However, if λ is one of the z_m, i.e $\lambda = z_m$, for some $m \in [1, M]$, the rank of the matrix differences reduces by one to M -1. This implies that we can find poles (z_m) as the generalized eigen values of the matrix pair $[X_0, X_1]$, i.e.,

$$X_1 q - \lambda X_0 q$$

Note that q, the generalized eigenvector, has no relationship to the eigen vectors of the correlation matrix. The M generalised eigen values of this matrix pair form the estimates of the z_m and the DoA may be obtained as

$$\phi_m = cos^{-1} \left[\frac{[\xi \ln(z_m)]}{kd} \right], m = 1, \ldots \ldots M \qquad (7)$$

The steps of Matrix pencil are therefore

1. Given N and M, choose L
2. Form matrices X and X
3. Find z_m as the generalized eigen values of the matrix pair $[X_0, X_1]$.
4. Find the DoA as specified as 10.

Note that finding the generalized eigenvalues of the matrix pair $[X_0, X_1]$ is equivalent to finding the eigenvalues of $[X_0^H X_0]^{-1} X_0^H X_1$.

Now after seeing the above theorem the similarities between matrix pencil and ESPRIT estimation techniques are clear. Both these algorithms estimates the diagonal matrix whose entries are poles of the system (what we call z_m). But apart from this similarlity, the major difference between this two techniques is that ESPRIT works with the signal subsplace as defined by the correlation matrix, but the matrix pencil works with the data directly. This represents a savings in terms of computation load.

The below Figure 6 shows the simulation analysis to estimate the accuracy of the DoA in the chosen matrix pencil theorem. The simulation was done using NS-3 simulator. In this DoA accuracy estimation simulation we have taken 5 snapshots and estimated in 7x7 correlation matrix. Since we have to locate only one signal effectively from the group member we have chosen the above criterion and the other signals which is emanated from the group member through multipath etc can be suppressed due to spreading gain. The assumption of 5 snapshot is adequate to estimate the signal subspace. In matrix pencil theorem the algorithm should have the knowledge about the data that is transmitted and also about the coherent detector. Since in our case the signal model is same and the type of the data which will be received will also be same and by this the signal to noise ratio is improved by averaging the received data. Another advantage of using the matrix pencil theorem is the computational loads are less and it is twice as fast as the other DoA estimation techniques.

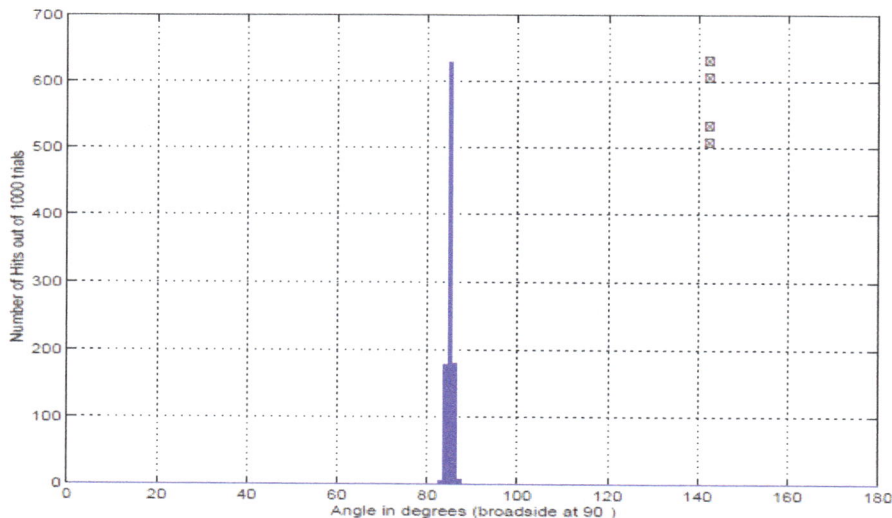

Figure 6. Accuracy estimation simulation for Matrix pencil theorem

4.1.2. Signal model and calculation of signal strength

For mathematical consideration we will assume that the signals experiences flat fading Assuming it as flat fading is a trade-off. Usually in the city condition the fading characteristics will change very quickly but at the same time we have to consider the ping pong effect in the city condition. As a result of this ping pong effect the group leader can change very fast and the group members also changes accordingly. So to incorporate this fast changing nature of group leader we would assume that the signals experiences flat fading. In an environment in which the fading characteristics are rapidly changing, this may not be valid. More importantly, estimating the correlation matrix is computationally intensive. So that's why we recommend to use Matrix pencil theorem to calculate the DoA. The signal model can be represented as below eqn (8).

$$X_i(t) = \sum_{k=1}^{dn-1} a(\theta_k) S_k(t - \tau_{ik}) + n_i(t)$$

(8)

$a(\theta_k)$ is the gain pattern of the receiver at the angle.

$n_i(t)$ is the additive noise.

d_{n-1} time delay that source k takes to travel from one mobile to another mobile.

τ_{ik} Is the number of mobiles transmitting signal to each other.

And the modulating signal can be represented as eqn (9)

$$S_k(t) = {}^1 g_k(t)\cos(w_0 t + \infty_k(t))$$

(9)

$$Z_1 = \begin{bmatrix} z_1 & z_2 & \cdots & z_M \\ \vdots & \vdots & \ddots & \vdots \\ z_1^{(N-L-1)} & z_2^{(N-L-1)} & \cdots & z_M^{(N-L-1)} \end{bmatrix}_{(N-L)\times M}$$

order. The group leader calculates LTE cell. The calculation is based on from the LTE as 100th percentile. All based on this each VNs will perform d. Whichever VNs has the highest l on the below equation (10).

(10)

$$Z_2 = \begin{bmatrix} 1 & z_1 & \cdots & z_1^{L-1} \\ 1 & z_2 & \cdots & z_2^{L-1} \\ \vdots & \vdots & \ddots & \vdots \\ 1 & z_M & \cdots & z_M^{L-1} \end{bmatrix}_{M\times L}$$

$$P_{dbm} = \begin{bmatrix} \begin{pmatrix} 0 & \dfrac{(N-M)_m}{} & 0 \end{pmatrix}_{dbm} \\ 0 & z_2 & \cdots & N0_m \\ \vdots & \vdots & \ddots & \vdots \\ 0 & 0 & \cdots & z_M \end{bmatrix}_{M\times M} \times 100$$

of the vehicle by the following

$$\Phi = \begin{bmatrix} \alpha_1 & 0 & \cdots & 0 \\ 0 & z_2 & & \\ \vdots & \vdots & \ddots & \vdots \\ 0 & 0 & \cdots & z_M \end{bmatrix}_{M\times M}$$

$$A \dfrac{P_{pollIteration}(t) \begin{bmatrix} \alpha_1 & 0 & \cdots & 0 \\ 0 & \begin{pmatrix}\dfrac{(N_M - M)_{dbm}}{N_M}\end{pmatrix} X100 - D_M \\ \vdots & \vdots & \ddots & \vdots \\ 0 & 0 & \cdot(d_1 \overline{\alpha_M} d_2) \end{bmatrix}_{M\times M}} = V_M$$

(11)

$P_{pollIteration}(t)$ Denotes the number polling done the system at a constant interval of time.

N_M Denotes the total number of members present in the group

M_{dbm} Denotes the signal strength received by the member from the LTE cell

$d_1 - d_2$ Denotes the distance travelled by the member from point D1 to D2.

V_M Denotes the velocity in the member node travels

5. SIMULATION RESULTS

The below simulations shows the comparison between the collision percentage and re-transmission attempts for different number of subscribers before and after applying the proposed algorithm. The simulations was carried using NS-3 simulator[7]. The results shows that definitely there is a marked difference in the collision percentage and re-transmission attempts after applying the proposed algorithm. Figure 7 and 8 shows the simulations for preamble ID collision percentage and successful preamble ID throughout condition. Before applying the algorithm the collision percentage for 10 subscribers accessing the RACH resource at the same time in a cell is around 20% but after applying the group based algorithm the preamble ID collision percentage for 10 subscribers is almost 11%. The algorithm has improved situation by reducing almost 9% of preamble ID collision. The advantage of this algorithm is more visible when the number subscribers increases. When 200 subscribers are attempting for the RACH the same time the preamble ID collision is around 75% but for the same number of subscribers after implementing the algorithm is around 60% which is 15 % less. Simulations in Figure 9 and 10 shows between the max-retransmission attempts while the collision percentage increases before and after applying the algorithm. Before applying the algorithm for the collision percentage of 20% the re-transmission attempt is 3. But after applying the algorithm for the same collision percentage of 13 % the re-transmission attempt is 1. So this shows that the number of re-transmission attempts and the collision percentage of the preamble ID has improved the situation after applying the proposed algorithm.

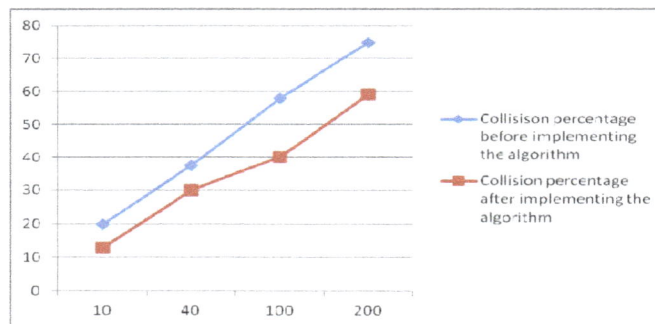

Figure 7. Preamble ID collision simulation results before and after applying algorithm

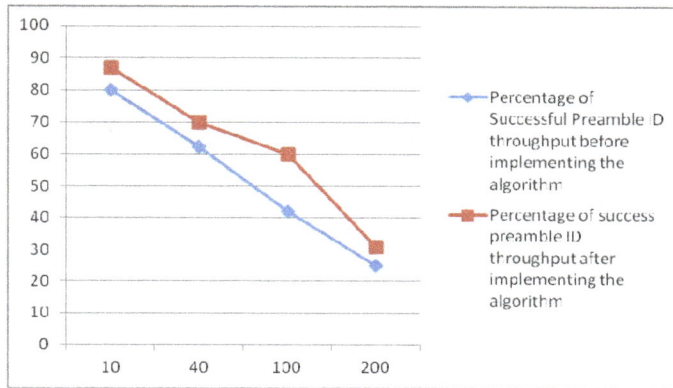

Figure 8. Preamble ID through before and after applying the algorithm

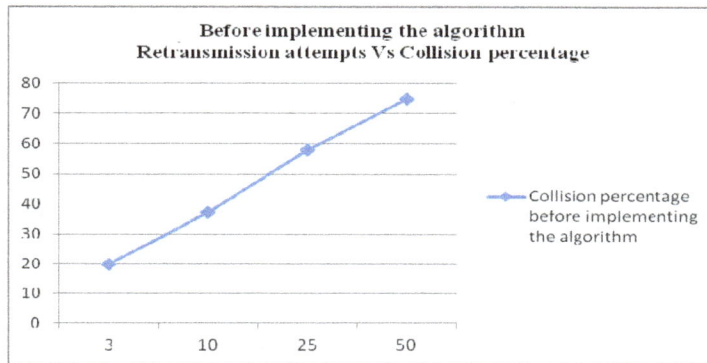

Figure 9. Retransmission attempts and collision percentage before applying the algorithm

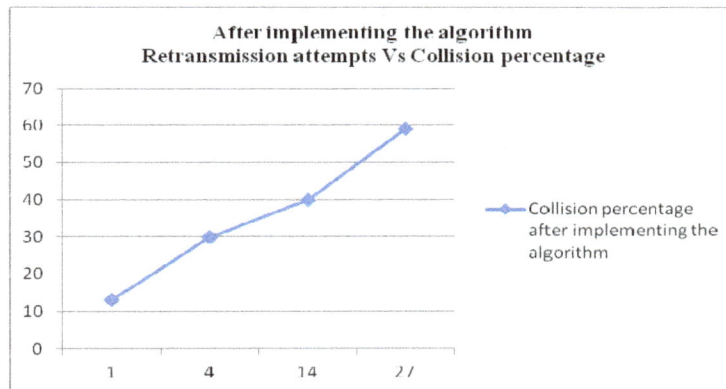

Figure 10. Retransmission attempts and collision percentage after applying the algorithm

6. CONCLUSION

The simulation is performed to analyze the preamble ID collision. During the contention based random access if VNs transmits same random preamble ID in different resource blocks there will not be any collision of preamble IDs but the collision occurs if VNs transmits the same preamble ID in the same resource blocks and this results in re-transmission of preamble ID automatically until the maximum re-transmission attempts is reached. As such the networks are busy because of the increasing customer base in LTE and as projected by ETSI many more machine to machine devices like smart meters, VNs, smart grids application devices etc are going to join the LTE bandwagon in the future. Hence an attempt has been made to improve the RACH congestion by the proposed algorithm and the results of the simulation of this algorithm is also encouraging in this regard. The strategy behind the proposed algorithm is to organise the access management mechanism of the machine to machine communication devices. Instead of allowing the VN devices to access the LTE base station at free will a group based access management techniques is introduced in the proposed algorithm. This not only organizes the access request sent by the VNs it will also avoid congestion in RACH and the impact of increasing VNs in the over H2H services can be reduced.

REFERENCES

[1]   Bureau of Infrastructure, Transport and Regional Economics -BITRE, (2009) "Greenhouse gas emissions from Australian transport: projections to 2020", Working paper 73, 2009, Canberra ACT.

[2]   Guoqiang Mao, "Responsive navigation and traffic control systems -The next generation in intelligent transport system design", CRIN Seminar, UTS Centre for Real-Time Information Networks, University of Technology Sydney, August 21, 2014.

[3]   Giuseppe Araniti, Claudia Campolo, Massimo Condoluci, Antonio Iera, and Antonella Molinaro, (2013) "LTE for Vehicular Networking: A Survey", IEEE Communications Magazine, May 2013, pp 148 - pp157.

[4]   3GPP TS 22.368 V11.3.0, (2011) "Service requirements for Machine-Type Communications (MTC)" Stage 1, September 2011.

[5]   Min Chen, Jiafu Wan and Fang Li, (2012) "Machine-to-Machine Communications: Architectures, Standards and Applications", Transactions on Internet and Information Systems, vol. 6, no. 2, February 2012.

[6]   ETSI, (2011) "Standards on Machine to Machine Communications", Mobile world congress, Barcelona.

[7]   S.-Y. Lien and K.-C. Chen, "Massive Access Management for QoS Guarantees in 3GPP Machine-to-Machine Communications," IEEE Commun. Letters, vol. 15, March 2011, pp. 311–13.

[8]   S.-Y. Lien, K.-C. Chen, and Y. Lin, "Toward Ubiquitous Massive Accesses in 3GPP Machine-to-Machine Communications," IEEE Commun. Mag., vol. 49, April 2011, pp. 66–74.

[9]   R. Paiva et al., "Overload Control Method for Synchronized MTC Traffic in GERAN," IEEE VTC-Fall, Sept. 2011, pp. 1–5.

[10]   J.-P. Cheng, C.-H. Lee, and T.-M. Lin, "Prioritized Random Access with Dynamic Access Barring for RAN Overload in 3GPP LTE-A Networks," IEEE GLOBECOM Wksps., Dec. 2011, pp. 368–72.

[11]   S.-T. Sheu et al., "Self-Adaptive Persistent Contention Scheme for Scheduling Based Machine Type Communications in LTE System," Int'l. Conf. Selected Topics in Mobile and Wireless Networking, July 2012, pp. 77–82.

[12]   A.-H. Tsai et al., "Overload Control for Machine Type Communications with Femto Cells," IEEE VTC-Fall, Sept. 2012, pp. 1–5.

[13]   NS-3: Simulator, http://www.nsnam.org/

OSC-MAC: DUTY CYCLE WITH MULTI HELPERS CT MODE WI-LEM TECHNOLOGY IN WIRELESS SENSOR NETWORKS

Mbida Mohamed and Ezzati Abdellah

Emerging Technologies Laboratory (VETE), Faculty of Sciences and Technology Hassan 1st University, Settat, MOROCCO

ABSTRACT

Recently, Wireless Sensor Networks (WSNs) grow to be one of the dominant technology trends; new needs are continuously emerging and demanding more complex constraints in a duty cycle, such as extend the life time communication . The MAC layer plays a crucial role in these networks; it controls the communication module and manages the medium sharing. In this work we use OSC-MAC tackles combining with the performance of cooperative transmission (CT) in multi-hop WSN and the Wi-Lem technology

KEYWORDS

WSN, MAC, Wi-Lem, CT, schedule, Duty Cycle.

1. INTRODUCTION

The Protocol OSC-MAC (On-demand Scheduling Cooperative Mac) is a protocol with a planning and request in order to solve the problems of energy of nodes transmitter, and gives as solution the integration of the technique of cooperation with the nearby nodes. In our work we integrate the Wi-Lem technology in order to delegate the nodes which have the higher energy and can be into the cooperation with the source, this solution brings an optimality of energy because the cooperative selection of nodes is applied in the Base station Wi-Lem, so we opted for 2 or more nodes which can be into the CT (cooperative transmission) according to the number of packets sending.

2. THE OSC-MAC PROTOCOL

As we know the life time of batteries is limited , what causes during the sending an imbalance energy in nearby nodes , and results an exhaustion power of the TRN (Transmitter root node) , to solve this problem we opt to initialize the technique by a phase of wakeup and decentralize the responsibility of sending's on the nearby nodes . Then according to the conception of nearby nodes which want to cooperate with these neighbour's and the others which refuse, during this work we are going to explain the state of art the request list of the cooperative nodes for a transmitter and the algorithm which engenders sending of packages by basing itself on the CT (cooperative transmission) RDV . The cooperating nodes are neither on the same duty cycle nor are they in the same collision domain. We use orthogonal and pipelined duty-cycle scheduling, in part to reduce traffic contention, and devise a reservation-based wake-up scheme to bring cooperating nodes into temporary synchrony to support CT range extension.

The major sources of energy consumption inherent to MAC's include idle listening, overhearing and collision, besides data transmission and reception. Idle listening means that nodes keep listening to the channel while there are no incoming packets at all - a case that has not been taken care of in many MAC protocols such as IEEE 802.11 where in WIFI stations must listen for possible traffic. Notably, idle listening is disastrous in WSNs based on the fact that nodes in this mode consume the same magnitude of power as in receiving. Overhearing means that nodes decode packets that are destined to others. Collisions result in corrupted packets and the following MAC layer retransmissions consume extra energy. From the network perspective, albeit these factors taper off individual node's lifetime, the network lifetime is more critically limited by the energy holes formed around the sink leaving unused energy outside of the holes. Many authors have considered duty-cycle MAC protocols, which allow nodes to alternate between active and sleep modes. These protocols dramatically reduce the periods of idle listening and overhearing.

3. SELECTION OF NODES IN ENERGY FACTOR BY USING THE WI-LEM TECHNOLOGY

Instead of making a selection of nearby nodes in the transmitter which results an additional loss of energy, for that reason we migrate this operation in a Wi-Lem station with large domain of communication radio.

3.1. WI-Lem technology

LEM has just create the family of Wi-LEM (Wireless Local Energy Meter, wireless local meters of energy) to allow the measure and the remote surveillance of the energy consumption as the power of batteries nodes or the water as well as the temperature and the humidity. With these meters, the industrial and tertiary companies can reduce their consumption of energy and water as well as identify the points of improvement. On the whole range Wi-LEM, allowing a large domain of communication radio between the nodes of the network, compared to the previous generation.

3.2 The advantages of WI –Lem

Wi-LEM presents additional benefits compared to the traditional divisional meters in wsn:

- At the level of the electric cupboard

- At the level of the installation of the network and its exploitation.

The following figure (1) presents the operating Wi-Lem CT technology in wsn:

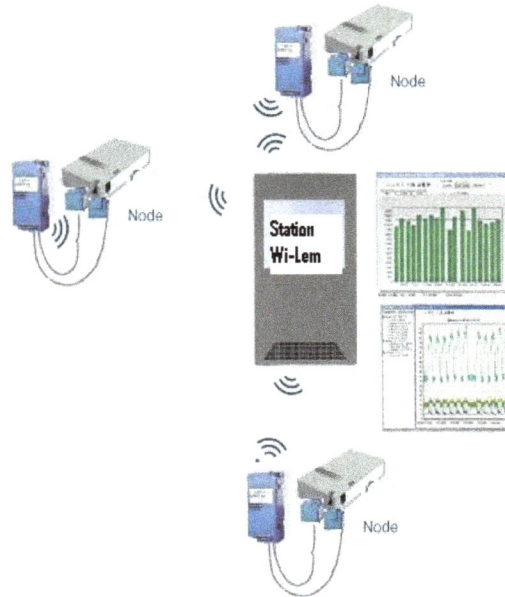

4. SELECTION OF THE NODES COOPERATION BY THE WI-LEM STATION

When a transmitter node wants to send its packets, first it transmit a packet request list of CTs nodes to the station Wi-Lem   by using an algorithm of selection nearby nodes possible to operate in CTs periods , the algorithm (1) and the figure (2) illustrates this function :

Algorithm 1 : allocation of the CTs in the nearby nodes

Input : List of the nearby knots organized has the order decreasing by percentage of energy : L
     The number of (octets) sent per packet : S
     Distance between transmitter and the next hop : D
Output : List of nearby nodes elected to transmit in CT phase   :
     LCT

For each li ∈ L do
     If Eli >= (Elect * S) + (Empl * S * D²) ;
     LCT ←li ;
     Else Li ← li+1 ;
     End
     Send LCt () ;
     End

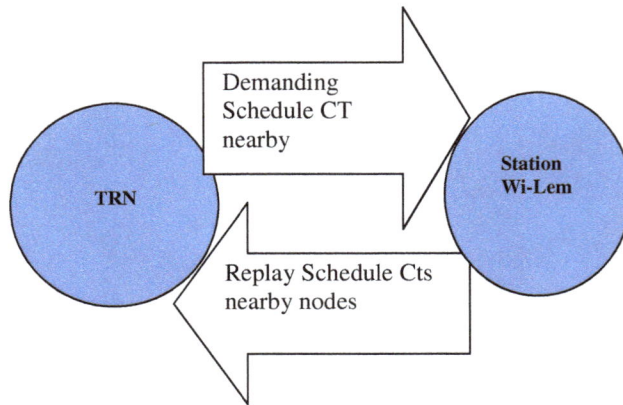

Figure 2: Diagram of the request /replay List CTs nearby

5. ELECTION OF THE NEARBY NODES WHICH WILL GO INTO THE CT MODE

Before the transmitter synchronizes its wakeup with the receiver as well as the nodes which want to run in CT Mode, the sending of packets is made according to report energy of nodes CT : { E ni ct (wi-lem) / i $\in$ {1...n} , n : number of nearby node's in wsn } obtained by the measurement of energy nodes in station Wi-Lem, and also get the necessary energy to send N packets by node transmitter : { E niT / i $\in$ { 1...t} , t : t ith node transmitter } , for this reason we develop The following algorithm (2) which explains the calculation of the necessary energy to transmit one packets of L bits :

Algorithm 2 : Calculation of energy transmission packets for L bits E ni T
Input : Emp-data : amplifying energy   Efs   : free space energy Output : Energy of sending one packet with L bits
if d is distance and do=sqrt (efs/emp); then sending L bit data over d distance if d>=do ; Energy=energy-(l*Etx+l*emp*(d^4)); else Energy=energy-(l*Etx+l*efs*(d^2)); End

As a reference of the ratio energy of nodes and the energy of transmission packet of L bits determine the number of helpers which are going to operate into the CT mode: the algorithm (3) describes the process of delegation nearby nodes CT.

Algorithm 3 : Selection of Elected nearby node CT
Input : Lists of the energy nodes which can enter in the CTs   RDV : E L (ct)   Number of packets transmission : N Output : List of elected nearby nodes : L ct-elu
For each ni ∈ E L (ct) do If E ni Ct / (N*E ni T) >= 1 L Ct-elu ←ni ; ni ←ni +1 ; Else ni ←ni +1 ; Send L ct-elu() ; End

6. ALLOCATION OF THE LISTS TRANSMISSIONS / RECEPTIONS PACKETS

In order to save the energy of every node sender we use a technique of Cooperative transmission which shows itself by the use of the nearby nodes according to its energy levels which run in routing mode of packets to the next hop, until the final Receiver. The transmission of packets are initialized by a phase of wake up and allocation of the interval of sending's which divides in no CT and CT intervals , in order to have   the transmitter node   and the cooperative nodes (helpers) are synchronized with the FR (final receiver) , the figure 3 illustrate   a description of this technique :

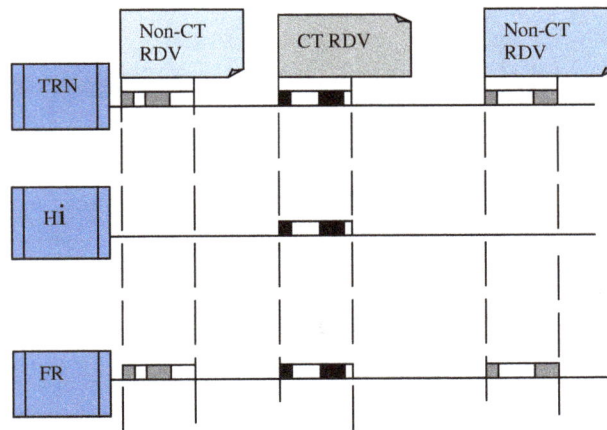

☐   Free period schedule
▢   CT RDV period schedule
■   No-CT RDV period schedule

Hi (the helper's) {the cooperative list nearby nodes / i ∈ {1 ... N}}

Figure 3: Diagram of allocation slot's time wake up for the CT and No-CT packets

The transmitter send a Super frame which defines the periods CT helpers and receiver in order to maintain the period of wake up, and to forward packets in the next hop, the leader Helper with the energy raised will send back a acknowledgement (ack) to inform the root transmitter node that the CT's meeting is reserved, the figure 4 describes this operative function. In no-Ct mode the sender node send directly a request in the next hop, up to the final receiver by indicating that it is going to transmit in a specific interval of time and the receiver will answer by a replay ack to inform it that the reservation of interval time is accepted or rejected.

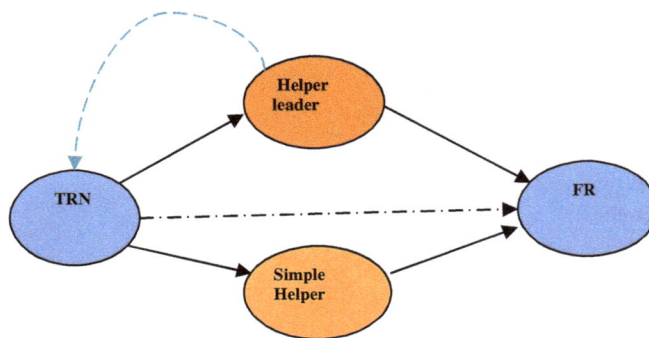

Figure 4: Diagram of transmission OSC-Mac by using CT Wi-Lem technology for 2 helpers

7. CONCLUSION

This paper propose the OSC-Mac Wi-Lem under exploitation of the cooperative transmission in goal as a technology to extend the life time of nodes by migrating the operations of classification energy's nearby nodes from the sender node to the base station Wi-Lem , and to list the nodes which are going probably to send into CTs periods .Indeed, it's used to choose the elected nearby nodes with higher power of the CTs period according to the number of packets sending, and simultaneously putting the other nearby nodes in a sleeping mode .

REFERENCES

[1] [1] IEEE Std 802.11b-1999 (R2003)," LAN/MAN Standards Committee of the IEEE Computer Society, IEEE-SA Standards Board, 2003.

[2] Jian Lin and Mary Ann Ingram School of Electrical and Computer Engineering Georgia Institute of Technology, Atlanta, Georgia 30332–0250.

[3] J. Li and P. Mohapatra, "An analytical model for the energy whole problem in many-to-one sensor networks," in Proc. IEEE VTC 2005I.

[4] J. W. Jung and M. A. Ingram, "Residual-Energy-Activated Cooperative

[5] Transmission (REACT) to Avoid the Energy Hole," in Proc. IEEE Int Communications Workshops (ICC) Conf, 2010, pp. 1–5.

[6] Y. Sun, S. Du, O. Gurewitz, and D. B. Johnson, "DW-MAC: a low latency, energy efficient demand-wakeup MAC protocol for wireless sensor networks," in Proc. MobiHoc 2008. New York, NY, USA: ACM.

[7] Synetica . Hilton House. High Street, Stone. Staffs. ST15 8AU U.K 2011.

[8] Salehdine Kabou sensor network energy, Bechar University, Algérie - Licence informatics 2010.

14

ZIGBEE BASED VOICE CONTROLLED WIRELESS SMART HOME SYSTEM

Thoraya Obaid, Haliemah Rashed, Ali Abu El Nour, Muhammad Rehan, Mussab Muhammad Saleh, and Mohammed Tarique

Department of Electrical Engineering, Ajman University of Science and Technology, Fujairah, United Arab Emirates

ABSTRACT

In this paper a voice controlled wireless smart home system has been presented for elderly and disabled people. The proposed system has two main components namely (a) voice recognition system, and (b) wireless system. LabView software has been used to implement the voice recognition system. On the other hand, ZigBee wireless modules have been used to implement the wireless system. The main goal of this system is to control home appliances by using voice commands. The proposed system can recognize the voice commands, convert them into the required data format, and send the data through the wireless transmitter. Based on the received data at the wireless receiver associated with the appliances desired switching operations are performed. The proposed system is a low cost and low power system because ZigBee is used. Additionally the proposed system needs to be trained of voice command only once. Then the system can recognize the voice commands independent of vocabulary size, noise, and speaker characteristics (i.e., accent).

KEYWORDS

Wireless, smart home, voice recognition, ZigBee, low power, low-cost, LabView,

1. INTRODUCTION

Home automation system has been around for more than a decade [1]. The main concept is to form a network connecting the electrical and electronic appliances in a house. This is a growing technology, which has changed the way people live. According to the data published by the market research and market intelligence firm ABI about 4 million home automation systems were sold globally in 2013 [8]. It is also estimated by the same organization that 90 million homes worldwide will employ home automation systems by the end of 2017. There have been several commercial and research versions of smart home system introduced and built [2,3,4,5,6]. But, none of the versions has broken through the mainstream yet other than security systems [7]. Smart home systems have captured many disparate technologies so far and products have been in the market for more than one decade. Many companies have entered in this field including Google. Google has announced an ambitious project named Android@Home [15] for smart home platforms. Despite over a decade of disparate activity in the industry no company has yet succeeded to launch home automation as a popular technology. The reasons of this failure have been comprehensively studied and listed in [1,16,17]. Some of the reasons are as follows: (a) cost: the existing systems are expensive and are owned by rich family with large house and estates, (b) difficult to install: expert professionals are needed to install and configure the system, (c) difficult to use: the control interfaces have poor quality and are not user friendly, (d) vendor dependency:

need to use separate systems for different companies' appliances, (e) less functionality: most of the system can either monitor or control the functions, and (f) not customized: most of the systems are not customized with the needs of the users. In addition to the above mentioned applications there are also some other reasons including security issues and multi-user problems [18,19].

Wireless communication based home automation system has gained a high momentum for the last couple of years. Wireless communication reduces the complexity related to the installation and maintenance compared to its wired counterpart. A typical wireless home automation system comprises battery operated and low power wireless sensors and actuators. Bluetooth, WiFi, and ZigBee are the popular choice for the backbone of such systems. Wireless network based smart home systems have become very popular as they provide comfort, security, and safety. Moreover, they support remote monitoring facilities. The availability of cheap wireless sensors, actuators, and modules has reduced the gap between the luxury and mass market segmentations of home automation technologies. However, wireless home automation system has some limitations too. The hostile radio channel, resource limitation, and mobility impose challenges for wireless home automation systems. Despite these limitations several organization and companies have developed wireless home automation system for diversified applications namely light control, remote control, smart energy, remote care, security and safety [21]. Especially, this industry has changed drastically since the introduction of cheap computers and laptops. The user interfaces of the home automation systems are much cheaper and user friendly now [20]. A typical wireless home automation system should have to deal with the following constraints: (a) high node density, (b) multipath radio wave propagation, (c) high interference, (d) multihop end-to-end connectivity, (e) dynamic topology, (f) various traffic patterns, (g) internet connectivity, and (h) secured communication. The major challenge is to deal with these constraints by using nodes that not only have limited memory and processing power, but also have limited operating life.

A typical wireless home automation system consists of two main parts namely (a) communication protocols, and (b) the user interfaces. The communication protocols are used for getting data to and from the home appliances. The user interfaces are used for monitoring and controlling them. There have been many solutions proposed for wireless home automation industry in the past few years. Some of them include (i) Z-wave, (ii) Insteon, (iv) Waveins, and (v) IP based solutions. The Z-wave is a wireless solution developed by Sigma Design. This wireless protocol has been promoted by the Z-wave alliance [12]. The main purpose of the Z-wave is to ensure a reliable transmission of short messages from a control unit to one or more nodes in the networks. Insteon is a home automation solution developed by SmartLabs and promoted by the Insteon Alliance. The major feature of Insteon is that it defines mesh topology composed of radio frequency (RF) link and power line link. The nodes can be RF only or power line links only or can support both types of communication. Waveins is a low power wireless protocol developed for controlling and monitoring appliances in a home. It is currently managed and promoted by Wavenis Open Standard Alliance [22]. This protocol defines physical, link, and network layers. Wavenis services can be accessed from the upper layer through an application programming interface. The IP-Based solutions have been initiated by Low Power Wireless Personal Area Network Group of Internet Engineering Task Force (IETF). This working group has defined mechanism for transmission of IPv6 packets on top of IEEE 802.15.4 networks. These networks have been named as LowPAN. The LowPAN follows the mesh topology and a routing protocol is used for its operation. The work on LowPAN is still in its infancy level and it is predicted that it will be an emerging technology for wireless home automation system in future.

Recently, ZigBee based solutions have drawn considerable attentions in the wireless home automation industry. This technology was developed by the ZigBee Alliance for low-data rate and short-range applications. ZigBee was designed for a suite of high level communication protocols used to create personal area networks. The initial version of ZigBee was based on IEEE 802.15.4

standard. It operates in the 868 MHz, 915 MHz, and 2.4 GHz bands in Europe, North America, and worldwide respectively. The ZigBee protocol stacks composed of four layers namely physical layer, network layer, medium access control, and application layer. The physical layer and the medium access control layer are based on IEEE 802.15.4 standard. The ZigBee Alliance has defined the application layer and network layer. The ZigBee defines three roles for the devices namely (i) co-ordinator, (ii) router, and (iii) end device. The co-ordinator and router have more functionality compared to the end devices. The ZigBee end devices can transmit data over longer distances via the router devices. The network layer supports both addressing and routing for tree and mesh topologies. In tree topology the coordinator acts as root. In mesh topology routes are discovered and maintained on-demand. The Ad hoc On-Demand Distance Vector (AODV) [23] has been chosen as the routing protocol in ZigBee networks. Two routing strategies are used namely point-to-point, and many-to-one. There are several application profiles in ZigBee. One of the profiles is Home Automation Public Application Profile, which makes ZigBee a suitable technology for home automation systems [24]. This application profile defines device descriptions, commands, and attributes for ZigBee applications in residential and commercial environments. Some of the applications of ZigBee include residential and commercial lighting, HVAC, security, wireless light switches, electrical meters, traffic management systems, and other consumer and industrial equipment that require short-range wireless transfer of data at relatively low rate. Considering all the above mentioned advantages we have selected ZigBee.

Voice controlled home automation systems have drawn considerable attention in the recent years. Initially, home automation systems were designed for the people seeking luxury and sophisticated home. But, there was always a need to develop home automation system for the people with special needs like the elderly and the disabled. According to a report published by the World Health Organization (WHO) around 785 million people of 15 years and older live with disability [49]. Of these, the World Health Survey reports that 110 million people have significant difficulties in functioning. Another report published by the Population Division of United Nations show that about 10% of the World's population is older than 60 years and it is estimated that this figure will reach up to 21% by 2050 [50]. In order to assist the old people and the people with disability home automation technologies are adopting voice control or voice recognition techniques. The main idea is to control and monitor home appliances by using voice command. The motivation behind this work is also the same. The rest of the paper is organized as follows. Section 2 presents some related works. System model has been presented in Section 3. Section 4 contains system operation, circuit details, and implementations. Finally, this paper is concluded with Section 5.

2. RELATED WORKS

One of the early experimental works on the ZigBee based home automation system was presented in [25]. This system was capable of monitoring door and window, smoke, gas leak, and water flooding in a home from remote location. Some simple control systems such as operating a valve and sending signal to security network have also been associated with this application. A ZigBee based home network system to track a user has been proposed in [26]. This system periodically tracks a user by using three systems namely Indoor Positioning System (IPS-M), Indoor Positioning System Infrastructure (IPS-M), and Indoor Positioning System Gateway (IPS-G). In ZigBee based wireless home automation system a gateway is an important component. One of such gateway architecture has been proposed in [27] to interconnect Digital Living Network Alliance (DLNA) compliant home appliances and a ZigBee network. In a similar work [28], another type of gateway architecture has been proposed to connect a low-rate home work with the internet. A user can control the home appliances via internet from a remote location through this gateway. A ZigBee based power monitoring system (PMS) has been proposed in [29]. In addition to ZigBee wireless communication the PMS also utilizes Digital Signal Processing (DSP) and Web services. DSP is used for real-time power parameters computation and Web Services are used for the communication infrastructure among distributed systems across a network. The proposed system has been constructed and validated for the power management in a campus. Another power management system has been proposed in [30]. This system stores the measurement data of current and voltage of the electric outlets in an embedded board. It can detect any overload in the system and send signal to a circuit breaker to turn off the power. A real-time home security system has been proposed in [31]. The system is able to detect intruder in a home and send messages via GSM network. The system also can receive instruction from a remote location to control the house appliances. A wireless smart home system based on ZigBee has been introduced in [32]. The system composed of three main components (i) home server with GSM module; (ii) intelligent environment detection sensor modules, and (iii) intelligent home appliances. By using these modules a home can be monitored remotely and an alarm message can be sent to a remote location. Most of the wireless home automation systems are constrained by limited operating range. Such constraints can be overcome by using multi-hop communication system. This was the main motivation of the work presented in [33]. In this work a home control system (HCS) based on multi-hop mesh network is presented. Three different interfaces based on ZigBee, Bluetooth, and GPRS have been used in HCS. The hardware architecture and software protocol have also been proposed. Some techniques to save battery life and hence maximize operating life of such network have also been proposed in the same work. A ZigBee based embedded remote control system has been implemented in [35]. The main advantage of this system is that it provides the wireless communication capabilities on an embedded board rather than on a PC. This kind of embedded board has made system smaller in size and power efficient. The interference between the home appliances and the home network has been investigated in [36]. The authors have shown that the presence of the home appliances in a Personal Area Networks (PAN) environment significantly affect the performance of the home network system which also operates in the same frequency band. A ZigBee based home automation system and Wi-Fi network have been integrated in [37]. The home automation system has been implemented by using Texas Instrument's MCU device LM359B96 in [38]. Users can access the system by a dynamic webpage of LwIP TCP/IP protocol stack or GSM sms. The performances of a ZigBee based home automation system have been compared with those of other technology based (i.e., WiFi and Bluetooth) systems in [39]. The authors claimed that ZigBee based system has longer life compared to WiFi and Bluetooth based home automation system. An energy efficient wireless sensor network has been proposed for smart home in [40]. This system uses energy efficient sensors. This system is situation based self-adjusting to reduce the energy consumption in home automation system. An automatic embedded software generation framework has been proposed in [41] to create and evolve ZigBee applications. The main target is to enhance the quality of digital home living environments. This framework allows rapid deployment of the supporting software for energy control and sensing devices to monitor energy usage in a home at any time. The performances of ZigBee technology have been investigated in [42]. The performances of this system such as latency, received signal strength indicator, and round trip delay have been investigated in this work.

Voice control system for ZigBee based home automation has been introduced in [34]. Speaker independent automatic speech recognition technique has been used. A number of modes have been used for convenience namely button trigger mode, voice password trigger mode, and circle recognition mode. A user can use any of the modes depending upon the conditions. A low power voice control system for home automation system has been proposed in [43]. In this system ZigBee network receives voice command as input to an ARM9 controller, which converts the data into a required format to be used in the microcontroller. Finally, the system generates some control characters to switch ON/OFF the home appliances. A client server based voice control system for home automation has been presented in [44]. Voice command is captured by a client system and is sent to a server via Wi-Fi wireless network. The server system converts the voice command into a form that is used to control the home appliances. Microsoft Speech (SAPI) has been used in this work for implementing voice recognition system. Mobile based voice command control and monitoring system has been implemented in [45], in which artificial intelligence has been used for voice recognition system. A multi-layer feed forward neural network has been used. Another similar work has been presented in [46]. Two controlling methods have been proposed namely time and speech. For speech recognition Microsoft voice engine tool has been used. A wireless home automation system has been designed for physically challenged to control the home appliances in [47]. Fault identification system has also been integrated in this system to monitor the conditions of the appliances. To provide security RFID based authentication system has been used. The results presented therein show that the system can control up to 20 appliances and 40 voice commands can be recognized by this system. In another recent work [48] the authors implemented a ZigBee based home automation system that could recognize up to 1225 voice commands with a success rate of 80%.

In this work we also presented voice controlled based wireless home automation system for elderly and people with disability. Our work is different from other related works in the following ways. We used National Instruments' LabView software. We have the following advantages due to LabView. The user interface is easier to design and implement. The system can be remotely controlled by a mobile or a computer. The system can easily be extended to include more appliances. The system is easy-to- install and configurable. Unlike other related system no expertise skill is required to install and configure the system. Microsoft's rich voice recognition library has been included in our work. The system needs to be trained only one time. Once it is trained, the system can system can recognize the voice commands independent of vocabulary size, noise, speaker characteristics or accent. The data acquisition is performed by using DAQ provided by the National Instruments. The use of DAQ simplifies the hardware implementations of this proposed system compared to other existing similar systems.

3. SYSTEM MODEL

The system model of this work is shown in Figure1. The system model consists of the following basic components (i) automatic speech recognition system, (ii) control units, (iii) wireless system, and (iv) application and home appliances.

Automatic speech recognition (ASR) system can be defined as an independent and computer-driven transcription of spoken language that allows a computer to identify the spoken words captured from a microphone or telephone and convert it into written texts. The main components of an ASR are (i) a microphone, (ii) speech recognition software, (iii) a computer, and (iv) a sound card. The ultimate goal of ASR is to allow a computer to recognize in real-time, with 100% accuracy, all words that are spoken by any person, independent of vocabulary size, noise, speaker characteristics or accent. Through a speech recognition program/application, the computer is able to process words one says and turn them into text that is displayed on the screen. There have been many research activities on the speech recognition system. The fundamental reasons of these research activities are (i) accessibility for the deaf and hard of hearing, (ii) cost reduction through automation, and (iii) searchable text capability.

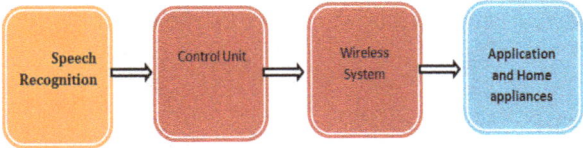

Figure 1. The system model

The control unit (CU) coordinates the components of a computer system. It fetches the code of all of the instructions in a program. It directs the operation of the other units by providing timing and control signals. All computer resources are managed by the CU. It directs the flow of data between the Central Processing Unit (CPU) and the other devices. There are different types of control program available including MatLab, C++, and LabView. In this work we choose LabView because of the following reasons: (a) it is easier to build a large program piecewise using small amounts of code, (b) it is easy to control the interfaces between hardware, (c) it is easy to find compatible hardware in the market, (d) it is simple to integrate with other hardware and software, and (d) it is easy to create the user-interface.

Wireless networks and sensors are playing important roles in emerging pervasive computing technologies that are required for the realization of smart homes. Effectively all wireless technologies that can support some form of remote data transfer, sensing, and control are candidates for inclusion in the smart home portfolio. In this work we used ZigBee wireless system. Compared to other wireless systems like Bluetooth and Wi-Fi, ZigBee has some advantages as follows: (a) ZigBee aims at automation whereas other two technologies aim at the connectivity of mobile devices in close proximity, (b) ZigBee uses low data rates, and consumes low power, (c) ZigBee networks support devices with longer range, and (d) ZigBee network is a robust network and that is easily scalable.

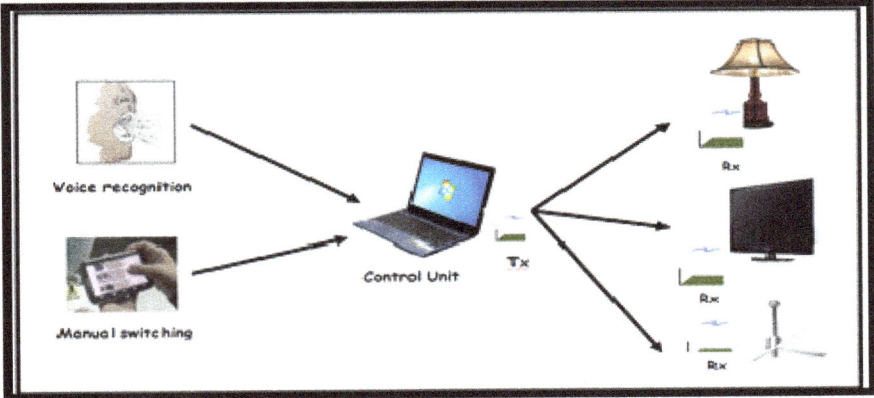

Figure 2. System operation

Applications and home appliances are the last step of the proposed system. Each device connected with the network must have a switching circuit to turn it On\ OFF. The status of each device is monitored by using toggle circuit. In addition to the switching devices a Data Acquisition Card (DAQ) has been used in this work. A typical commercial DAQ card contains ADC and DAC that allow input and output of analog and digital signals in addition to digital input/output channels. There are many types of DAQ cards are available to provide interface between the LabView software and the hardware. In this project NI USB 6008 DAQ card has been used.

4. SYSTEM OPERATION

The basic system operation of the proposed system is shown in Figure 2. A user has two options for controlling the home appliances (a) by using manual switching and (b) by using voice recognition. The control unit is an interface program that must satisfy the following two conditions (a) the output from the interface program is forwarded to a wireless transmitter and sent to a receiver through wireless channel, and (b) the receiver at the appliances accept the receive signal to turn ON or OFF the device. The front panel of the LabView program is shown in Figure 3. The system operation can be described based on this front panel in the following steps:

- The voice command is captured by using a microphone and sent to the computer.
- Voice recognition is performed by using LabView with the help of Microsoft's rich speech library.
- Upon recognition of the voice command control characters are generated and sent through ZigBee wireless communication protocol to specified application address.
- At receiver side, application can turn ON or OFF a relay controlling circuit depending on the received controlled characters.
- A hardware switching toggle circuit is added to simplify the switching process and to allow for ON/ OFF switching with a single word voice command.
- As an additional feature the system can be controlled using smart devices.

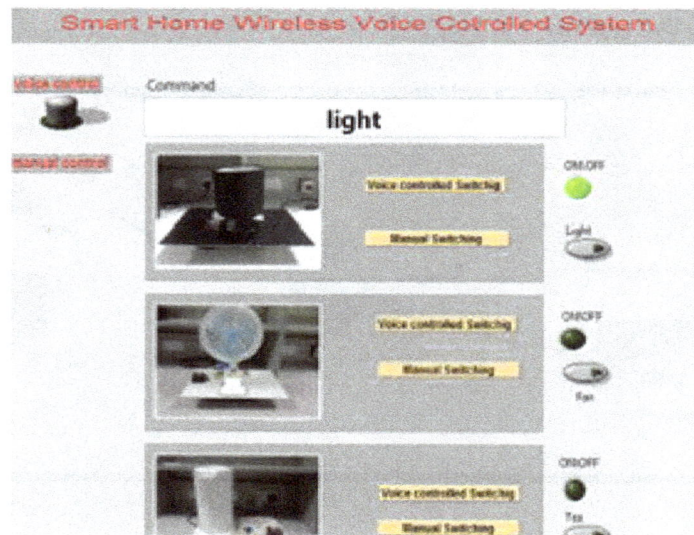

Figure 3.   The front panel of the program.

The circuit block diagram of the transmitter system is shown in Figure 4. The speech recognition is dealt with LabVIEW. So the program recognizes the speech with library inside it. Then the output from LabView is sent to the DAQ. The DAQ connects the microcontroller and transmitter.

The transmitter sends commands to the wireless receiver that is connected with the switching circuit.

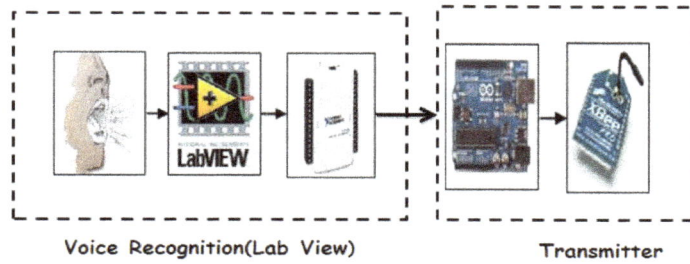

Figure 4. Transmitter Circuit.

The LabView takes input commands from the voice recognition system, compares the command with each stored command inside the library of LabView. The comparison is done in terms of the frequency. If the frequencies of the commands are matched, the LabView will generate an output pulse that is sent via DAQ. The detail operation of the transmitter is illustrated in Figure 5.

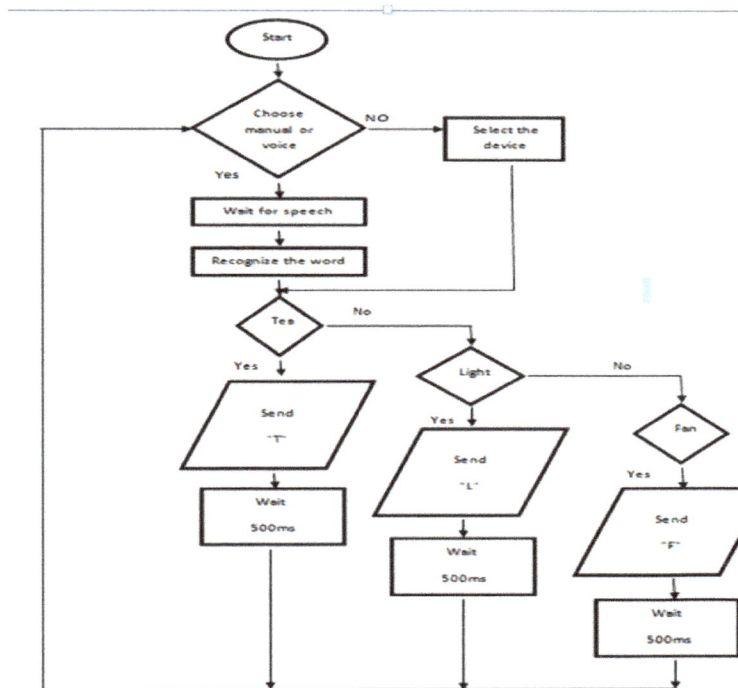

Figure 5 The flowchart for the operation of the transmitter.

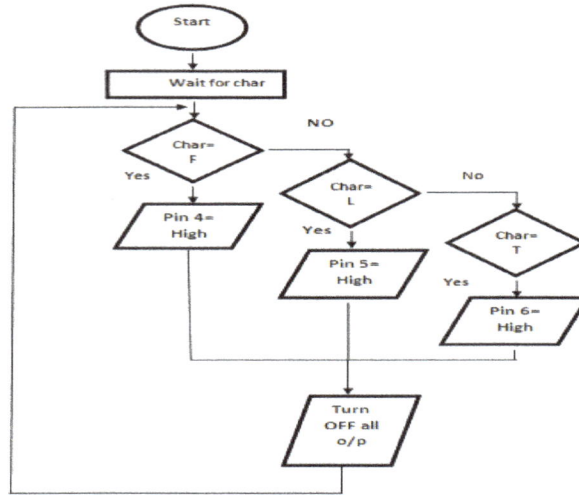

Figure 6 The flowchart for the operation of the receiver

The wireless transmission unit consists of Xbee and Arduino. The Arduino is programmed in such way that if it receives the pulse code 100 from labVIEW program, it will generate a character "T". If it receives the pulse code 010, it will generate a character "L". If it receives the pulse code 001, it will generate a character "F". Then the program will wait for 500ms after transmitting a new data. The receiver module is connected with the switching circuit of the appliances. We connect the pins of the receiver module to the home appliances as follows: Pin 4 to the fan, Pin 5 to the light, and Pin 6 to the tea machine. The program waits for receiving a character. For example, if the program receives character "F" it will turn ON the fan. If the program again receives the same character, it will turn OFF the fun by using the switching circuit. Similar procedure is followed for switching ON or OFF the other appliances (i.e., light and tea machine). Figure 6 shows the flowchart of the processes of wireless receiver.

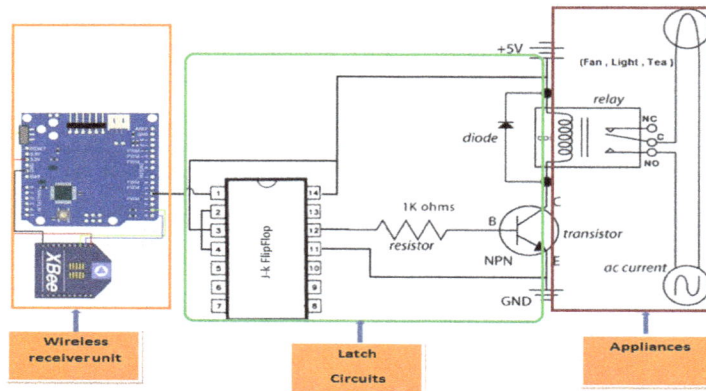

Figure 7. The latch circuit

In this work, we use the same word to turn ON/OFF the appliances. Since the LabView cannot turn ON/OFF the appliances by using the same word, we use a switching circuit to switch an appliance between the two states. It means that if a character is received for the first time, the program will turn on the device. If the same character is received for the second time, the program will turn off the device. We have designed a latch circuit for this type of switching operation. The latch circuit is connected with the relay that switches ON\OFF the device. The latch circuit is shown in Figure 7. The latch is a basic storage element in a sequential logic. Flip-flops and latches are the fundamental building blocks of digital electronics systems used in computers, communications, and many other types of systems. The latch circuit used in this work has been implemented by using a J-K flip-flop. In electronics, a flip-flop or latch is a circuit that has two stable states and can be used to store state information. The circuit can be made to change the state by applied signals. When used in a finite-state machine, the output and next state depend not only on its current input, but also on its current state (and hence previous inputs

5. CONCLUSIONS

In this work, we have implemented a voice controlled Zigbee-based home automation system. We used speech recognition system to implement this work. The LabView software has been used to implement the voice recognition system. The main advantage of the system is that it does require training of voices for only one time. At the same time LabView software has been used to support human-computer interactions to realize multiple functions. The system is designed for elderly and disabled people so that they can monitor and control the home appliances with their limited ability. The wireless part of the system has been implemented by using Zigbee RF modules. Hence, the system is highly efficient and it consumes low power. This system can be easily extended to remotely control the home appliances through smart devices like iPhone/ iPod and others phones so that one can remotely check the status of the home appliances and turn ON or OFF the same. The proposed system has been tested with three home appliances. But, it can be easily expanded to include more home appliances.

REFERENCES

[1]   A.J. Bernheim Brush, Bongshin Lee, Ratul Mahajan, Sharad Agrawal, Stefan Saroiu, Collin Dixon (2011), "Home Automation in the Wild: Challenges and Opportunities", Proceedings of ACM CHI Conference on Human Factors of Computing System, May 7-12, Vancouver, BC, Canada

[2]   Richard Harper (2003) "Inside the Smart Home", Springer-Verlag London Limited 2003

[3]   Juile A Kietz, Shwetak N. Patel, B rian Jones, Ed Price, Elizabeth D. Mynatt, Gregory A. Abowd (2008) "The Georgia Tech aware Home", The 26th CHI Conference, April 5-10, Florence, Italy ,2008, pp. 3675-3680

[4]   Tiiu Koskela and Kaosa-vaananen-vainio-Mattila (2004) " Evolution towards smart home environments:empirical evaluation of three users interfaces, International Journal of Personal and Ubiquitous Computing, July, Vol. 8, No. 3-4, pp. 234-240

[5]   Daine J. Cook and Sajal K. Das (2005) "Smart Home Environments: Technology, Protocols, and applications", Wiley series of Parallel and Distributed Computing (kindle Edition) , pp. 273-294

[6]   Tom Rodden and Steve Benford (2003) " The evolution of buildings and implications for the design of ubiquitous domestic environments", Proceedings of the SIGCHI Conference on Human Factors in Computing Systems", April 5-10, Fort Lauderdale, Florida , pp. 9-16

[7]   Home security systems, home security products, and home alarm systems available at http://www.adt.com

[8]   ABI Research on home automation available at http://www.abirearch.com

[9]   Vijay Saraswat, Bard Bloom, Igor Peshansky, Olivier Tardieu, and David Grove, " X10 Language Specifications Version 2.4", September (2013) available at http://x10.sourceforge.net/d ocumentation /languagespec/x10-240.pdf

[10] Insteon White Paper: The Details available at http://www.insteon.com/

[11] Y. Kyselytsya and Th. Weinzierl, " Implementation of the KNX Standard" available at http://www.weinzierl.de/

[12] The Z-Wave technology available at http://www.z-wavealliance.org/technology

[13] ZigBee Home Automation available at http://www.zigbee.org/Standards/

[14] IoBridge Connect available at http://connect.iobridge.com/docs/

[15] Janko Roettgers, "Time for Google's Android @Home to Make a New Splash" available at http://www.businessweek.com/articles/2013-05-08/

[16] W. Keith Edwards, Rebecaa E. Grinter, Ratul Mahajan, and David Wetherall (2011) " Advancing the state of home networking", Communication of ACM, Vo. 5, No.6, July, pp. 62-71

[17] Forest and Sullivan, "North American Home Automation Market", available at http://www.forest.com/sublib

[18] Intile S.S. (2002) "Designing a home of the future", IEEE Pervasive Computing, Vol. 1, No.2, pp. 76-82

[19] Scott Davidoff, Min Kyung Lee, Charles Yiu, John Zimmerman, and Anind K. Dey (2006) "Principles of smart home control", Proceedings of the 8th International Conference on Ubiquitous Computing, 17-21 September, pp. 19-34.

[20] Frost and Sullivan, " Innovation and Affordability open up home automation to alarge audience" available at http://www.frost.com/prod/servlet/press-release- pag?docid=273090581

[21] Carles Gomez and Joseph Paradells (2010) " Wireless Home Automation Networks: A survey of Architectural and Technologies", IEEE Communication Magzine, June (2010), pp. 92-101

[22] http://www.radiocomms.com.au/products/42985-Wavenis-Open-Standard-Alliance

[23] C. Perkins, E. Belding-Royer, and S. Das, "Ad hoc On-Demand Distance Vector (AODV) Routing" available at http://www.ietf.org/rfc/rfc3561.txt

[24] ZigBee Alliance, " ZigBee Home Automation Public Application Profile", version 25, vol. 1.0, October 2007.

[25] Dechuan Chen and Meifang Wabg (2006) " A home security ZigBee network for remote monitoring application", Proceedings of IET International Conference on Wireless, Mobile, and Multimedia networks", November 6-9, Hangzhou , China, pp. 1-4

[26] Woo-Choo Park and Myung-Hyun Yoon (2006) " The Implementation of Indoor Location System to Control ZigBee Home Network", Proceedings of International Joint Conference SICR-ICASE" October 18-21, Busan, South Korea, pp. 2158-2161

[27] Kawamoto, R., Emon , T., Sakata, S., and Youasa, K. (2007) " Energy efficient sensor control scheme for Home Networks based on DLNA-ZigBee Gateway Architecture", Proceedings of the First International Global Information Infrastructure Symposium, July 2-6, Marrakech, Morocco, pp. 73-79

[28] Zhang Shunyang, Xu Du, Jiang Yongping, Wang Riming, " Realization of Home Remote Control networks based on ZigBee (2007) " Proceedings of the 8th International Conference o Electronic Measurements and Instrumentations, August 16-18, Xiang, China, pp. 4-344-4-348

[29] Jui-Yu Chang, Tao-Yuan, Min-Hsing Hung, and Yen-Wei Chang (2007) " A ZigBee based Power Monitoring System with Direct Load Control Capabilities", Proceedings of IEEE International Conference on Networking, Sensing, and Control, April 15-17, London, pp. 895-900

[30] Ying-Wen Bai and Chi-Huang Hang (2008) " Remote Power ON/OFF control and current measurement for home electric outlets based on a low-power embedded board and ZigBee Communication", Proceedings of IEEE International Symposium on Consumer Electronics, April 14-16, Vilamoura, pp.1-4

[31] Jan Hou, Wu Cang dong, Zhongjia Yuan, Jiyuan Tan (2011), " Research of Intelligent Home Security Surveillance System based on ZigBee", Proceedings of the Initial Symposium on Intelligent Information Technology Application Workshops, December 21-22, Shanghai, pp. 554-557

[32] Jianfeng Wu, and Hubin Qin (2008) " The deisgn of wireless intelligent home system based on ZigBee", Proceedings of the 11th IEEE International Conference on Communication Technology, November 11-12, HagZhau, pp. 73-76

[33] Fei Ding, Guangming Song, Jianing Li, and Higuo Song (2008) "A ZigBee Based Mesh Network for Home Control", Proceedings of International Workshop on Geoscience and Remote Sensing", Shanghai, 2008, Vol. 1, pp. 744-740

[34] Jieming Zhu, Xuecai, Yucang Yang, and Hang Li (2010) " Developing a voice control system for ZigBee based home automation", Proceedings of IEEE International Conference on network Infrastructure and Digital Content, September 24-26, Beijing, 2010, pp. 7737-741

[35] Cui Chenguyi, Zhao Guannan, and Jin Mingle (2010) " A ZigBee based embedded remote control System, Proceedings of the 2nd International Conference on Signal Processing Systems, 5-7 July, 2010, Dalian, pp. v3-373-376

[36] Simek, M., Fuchs, M, Mraz, L., and Morvek, P. (2011) " Measurement of LowPAN Network co-existence with Home Microwave Appliances in Laboratory and Home Environments", Proceedings of International Conference on Broadband and Wireless Computing, October 26-28, 2011, Bercelona, pp. 292-299

[37] Ming Zhi Wu, Wei-Tsang Lee, Ren-JiLino, Chaye, G. (2012) " Development and validation of an integrated dynamic security monitoring platform," Proceedings of the 6th International Conference on Genetic and Evolutionary Computing, August 25-28, 2012, pp. 524-517, Kotakushu, 2012

[38] Chunglong Zhang, Min Zhang, Young Sheng Su, and Weillian Wang (2012), " Smart home design based on ZigBee wireless sensor network", Proceedings of the 7th International ICST Conference on Communications and networking in China, Kun Ming, pp. 463-466

[39] Rathod K., Parikh, N., and Shah, V. (2012) " Wireless automation using ZigBee protocols", Proceedings of the 9th International Conference on Wireless and Optical Communication, Indere, September 20-22, 2012, pp.1-5

[40] Jin Sung Byum, Boungju Jeon, Junyoung Noh, and Youngil Kim (2012) " An Intelligent self-adjusting sensor for smart home services based on ZigBee Communiation", IEEE Transaction on Consumer Eletronics, Vol. 58, No. 3, pp. 799-802

[41] Chihhsiong Shih and Bwo-cheng Liang (2012) " A model driven software framework for ZigBee based energy saving systems", Proceedings of the 3rd International Conference on Intelligent Systems, modeling and Simulation, February 8-10, Kota Kinabaiu, pp. 487-492

[42] Karia Deepak, Jaypal Bavisker, Raj Makwana and Panchat Niraj (2013) " Performance analysis of ZigBee based load control and power monitoring system", Proceedings of the International Conference on Advances in Computing, Communications, and Informatics, Myshore, August 22-25, 2013, pp. 1779-1484.

[43] Y.B. Krishna and S. Nagendram (2012) " ZigBee Based Voice Control System for Smart Home", International Journal on Computer Technology and Applications, Vol. 3, no. 1 (2012), pp. 163-168

[44] B. Mardiana, H. Hazura, S. Fauziyah, M. Zahariah, A.R. Hanim, and M. K. Noor Shahida (2009) " Home Appliances control Using Speech Recognition in Wireless Network Environment", Proceedings of the International Conference on Computer Technology and Development, pp. 285-288.

[45] N.P. Jawarkar, V. Ahmed, and R.D. Thakare (2007) " Remote Control using mobile through spoken commands", Proceedings of IEEE International Consortium of Stem Cell Networks", 2007, pp. 622-625.

[46] S.M.A. Haque, S.M. Kamruzzaman, and Md. Islam (2006) " A system for Smart Home Control of Appliances based on timer and Speech Interaction", Proceedings of the 4th International Conference on Electrical Engineering, January 26-28, pp. 128-13

[47] Gananasekar, A.K., Jayarelu, P., and V. Nagrajan (2012) " Speech recognition based wireless automation of home with fault identification for physically challenged" , Proceedings of International Conference on Communications and Signal Processing, April 4-5 2012, Chennai, pp. 128-132

[48] Al Shueili, H. and Sen Gupta G, Mukhopadhyaya (2001) " Voice recognition based wireless home automation system", Proceedings of the International Conference on Mechatronics , May 17-19,, Kualalumpur, pp. 1-6

[49] http://whqlibdoc.who.int/hq/2011/WHO_NMH_VIP_11.01_eng.pdf

[50] http://www.un.org/esa/population/publications/worldageing19502050/

15

ZIGBEE BASED WEARABLE REMOTE HEALTHCARE MONITORING SYSTEM FOR ELDERLY PATIENTS

Khalifa AlSharqi, Abdelrahim Abdelbari, Ali Abou-Elnour, and Mohammed Tarique

Department of Electrical Engineering, Ajman University of Science and Technology, Fujairah, United Arab Emirates

ABSTRACT

Remote health care monitoring system (RHCMS) has drawn considerable attentions for the last decade. As the aging population are increasing and at the same time the health care cost is skyrocketing there has been a need to monitor a patient from a remote location. Moreover, many people of the World are out of the reach of existing healthcare systems. To solve these problems many research and commercial versions of RHCMS have been proposed and implemented till now. In these systems the performance was the main issue in order to accurately measure, record, and analyze patients' data. With the ascent of wireless network RHCMS can be widely deployed to monitor the health condition of a patient inside and outside of the hospitals. In this work we present a ZigBee based wireless healthcare monitoring system that can provide real time online information about the health condition of a patient. The proposed system is able to send alarming messages to the healthcare professional about the patient's critical condition. In addition the proposed system can send reports to a patient monitoring system, which can be used by the healthcare professionals to make necessary medical advices from anywhere of the World at any time.

KEYWORDS

ZigBee, Wireless, patient, monitoring, healthcare, wearable, sensor, networks, data publishing

1. INTRODUCTION

Over the recent years remote health care monitoring systems for the elderly people have drawn considerable attentions. According to UNFPA, the global population is no longer young for the first time in the history [1]. Population ageing is affecting the entire world and is happening in all regions. But, it is progressing at a faster rate in the developing countries. Seven out of the fifteen countries in the developing world have more than 10 million old people. By the year 2050 another fifteen developing countries are expected to have 10 million old people. It is worthwhile to mention here that the average life expectancy in the United States was 47.3 years in 1900. But, it has increased to 68.2 years and 77.3 years in 1950 and 2002 respectively [2,3]. People are living longer because of better nutrition, sanitation, medical advances, education, economic well-being, and health care. Population ageing poses challenges to individuals, families, and societies. By adopting proper policies societies should be prepared for an ageing world. Overall, the older people should not be considered as a burden for the society. Their wisdom, energy, and experience are added advantages for us to take care of the challenges of the 21st century. In order to keep the ageing population healthy we have to deal with some challenges. The major challenge for us is to keep them healthy with our limited resources. Although numerous groundbreaking achievements have been noticed in the health care sector for the recent years, the health care expense is still sky high and it has become an issue that even the developed countries are worry about. According to the data provided by the Kaiser Family Foundation [4] the per capita expenditure of the health care is increasing at an exponential rate in some countries as shown in Figure 1. With such a high and continuously increasing healthcare expenses, medical care for the ageing people is becoming progressively challenging. One of the reasons for this high medical expense is hospitalization cost. The senior citizens are the most frequent visitors to the hospitals. They visit the hospitals for their medical treatment. Sometimes they have to stay there for a certain period of time for follow up of their medical treatment. Their staying in a hospital not only incurs expenses, but also incurs loss of patient's mobility. Remote Health Care Monitoring System (RHCMS) has been proposed as a solution to this problem. The main concept is to monitor a patient from a remote location and to provide her/him with necessary medical advices. The RHCMS has numerous advantages compared to conventional healthcare systems. Some of the advantages include (a) monitoring a patient, (b) responding to an emergency, (c) assisting patient mobility, (d) shortening hospital stay, and (e) reducing medical expenses. By using RHCMS the physical conditions of the patients can be monitored for twenty hours a day and seven days a week. The emergency services can be provided to the patients with a minimum delay. The patients can be served without going to a health care facility and admitting there. The healthcare professional can perform the follow up from a remote location and hence a patient needs to stay in a hospital for a short period of time. In a nutshell RHCMS reduces expenses related to the medical services.

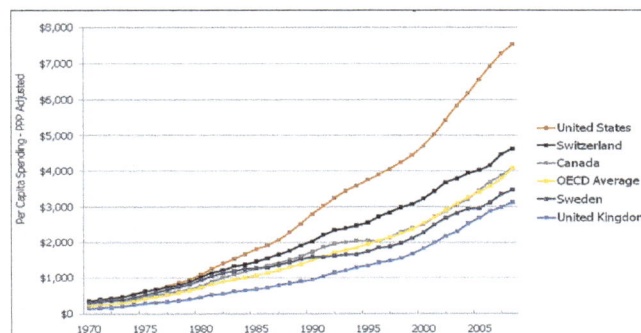

Figure 1 Per capita health acre expenditure of the World

Most of the proposed RHCMSs are based on wireless technology. The evolution of the wireless network has experienced a very fast-paced. Since the introduction of the IEEE 802.11 protocol wireless networks have experienced a huge market demand. Within the four years of the introduction wireless networks became very popular because of its portability, convenience, ease of installation, and low cost [5]. In that time period about 7.5 million households in the United States deployed wireless network. Wireless networks and medical sensors have been combined in RHCMS. There have been many medical sensors available in the market. Most of the sensors can measure and display critical health monitoring data such as the pulse rate, blood pressure, temperature, and blood sugar of a patient. One of the limitations of the proposed RHCMS is its limited coverage area. The measured data cannot be transmitted beyond a certain distance. Thus, it is not possible for the healthcare professionals to monitor the medical conditions of a patient from a distant location. In a hospital either the nurse or the physician has to move from one patient to another patient for monitoring them. Hence, it may not be possible for them to monitor a patient's health conditions all the time. This situation can be even worse when they have to take care of a large number of patients in a hospital at a given time. In order to overcome the above mentioned limitations an on-line health monitoring system has been proposed in this work. The proposed system can monitor the temperature, pulse, muscle, and ECG data of a patient. The proposed system has been designed by using the ZigBee technology. A major portion of this system has been implemented in LabView. Hence, the proposed system is reconfigurable as per users' need. The proposed system has been tested and verified in order to ensure its accuracy and reliability. The system consumes a very low power. The system consumes a very low power because it transmits signal only if the monitored parameters (i.e., temperature, heart beat rate etc.) go outside their normal ranges. Otherwise, the system puts the transmitter into a sleep mode to save energy. Hence the proposed system has a long operating life. The rest of the paper is organized as follows. Section 2 presents some related work. The system model of the proposed RHCMS has been presented in section 3. The implementation and results have been presented in section 4. This paper is concluded with section 5.

2. RELATED WORKS

Numerous prototypes for remote health care system can be found in the literatures. Since this work is based on the ZigBee technology, we focus only on the remote health care systems that have been designed based on the ZigBee technology.

One of the early works on health care monitoring system has been proposed in [6]. The proposed system is suitable for patients, senior citizens, and others who need continuous monitoring of their health. The proposed system can monitor the ECG signals of a patient based on Session Initiation Protocol (SIP) and a ZigBee network. The system consists of a wireless ECG sensor, ECG console, ZigBee module, SIP register, a proxy server, a database server, and wireless devices. Simultaneous monitoring of the biomedical signals from multiple patients has been addressed in [7]. The proposed network is based on IEEE 802.15.4 standard and the ZigBee technology. The authors have proposed an optimized source routing protocol to control the network load. Some other issues including energy consumption, network lifetime, and delivery ratio have also been addressed in the same work.

An intelligent remote healthcare system based on power line communication and the ZigBee network has been proposed in [8]. The system consists of physiological sensors, a ZigBee/PLC gateway, and some special software. The physiological data are collected by the physiological sensor and are sent to a controlling center through a ZigBee/PLC gateway. The data are stored and analysed at the controlling center.

A low power microcontroller based patient bed monitoring system has been presented in [9]. Resistive bend sensor has been used for minimizing the harmful effects of bedsores, which is a common problem in hospital's intensive care units and assistive living environments during rehabilitations. The proposed system is able to replace the current wired system and it supports continuous patient monitoring by enabling the patient bed mobility.

A Wireless Body Area Sensor network (WBASN) based on the ZigBee technology has been presented in [10]. All nodes in the WBASN are connected as star topology and central node (i.e., access point) is used to control the network. The data collected by the access point is transferred to a hospital network over mobile communication network. The authors have proposed a novel "wake-up on-demand" mode of network operation. According to this mode the network "sleeps" when the most of the circuits are turned off to reduce power consumption. Once the WBASN is waked up all modules begin to work and all biomedical signals are obtained, stored, and transmitted.

Reliability of data transmission for healthcare monitoring system has been investigated in [11]. The authors have suggested that there is always a chance for loosing physiological data when a number of ZigBee devices operate in a hospital at a given time. Although the medium access control is taken care of by the MAC layer, the authors have designed and implemented a new medium access control algorithm to ensure reliable data transmission of the physiological data.

A telemedicine information monitoring system consists of vital sign monitoring devices, a healthcare gateway, and a health service information platform has been proposed in [12]. Among these components the healthcare gateway is the most critical component. The ZigBee module is used to transmit information between the vital sign monitoring devices and the healthcare gateway. The vital sign monitoring devices include ECG, SPO2, blood pressure, glucose, and body temperature. The data is then relayed to a healthcare service information platform. The system is based on Service Oriented Architecture (SOA) concept to provide the healthcare management for people who are suffering from chronic illness.

A remote patient monitoring system based on the ZigBee wireless sensor network and the Internet Things has been introduced in [13]. The system generates electronic medical records that are saved in a database. After analysing the data the proposed system can send feedback about the diagnosis, medical programs, and proposals to a remote location. The system uses the ZigBee network for real time transmission of the physical data. The data processing and information releasing have been implemented by a database program.

A wearable remote healthcare system for assessing hydration status and visceral fat accumulation by using Bioelectrical Impedance (BI) analysis has been proposed in [14]. The authors have designed a ZigBee based BI to replace the conventional wired BI. The proposed system consists of BI measurement circuit integrated with 0.35 μm CMOS technology and a transducer circuit of the ZigBee module.

Two alternative systems have been proposed for the deployment of the ZigBee based wireless personal area network (WPANS) for remote patient monitoring in the general wards of a hospital in [15]. In the first approach a single WPAN is considered for gathering and transmitting physiological data from the patients in a ward. In the second approach multiple WPANS are considered. The simulation results show that the multiple WPANS out-perform the single one in respect of efficiency and reliability for data transmission.

An expandable wireless health monitoring system based on ZigBee has been proposed in [16]. The proposed system can monitor the temperature and pulses of a patient wirelessly. The test result presented therein shows that the proposed system can monitor the temperature and pulse of a patient with a high accuracy.

An ambient care system (ACS) framework to provide remote monitoring, emergency detection, activity logging, and personal notification services has been proposed in [17]. The proposed system consists of Crossbow MICAZ devices, sensors, and PDA enabled with ZigBee technology. The authors concluded that the combination of the ZigBee technology together with a service oriented architecture is the best option for ACS services.

A wireless sensor gateway (WSG) has been proposed in [18] to monitor patients' health. The main objective of this project is to monitor the cardiovascular status of a patient. Biological signals like ECG, pulse wave, and body weight are the important parameters for the cardiovascular monitoring of a patient. The proposed gateway is deigned to receive data from wireless sensors through a ZigBee interface and to forward the same to a personal computer via Bluetooth interface.

A ZigBee based system for remote monitoring of SPO_2 has been proposed in [19]. The system consists of SPO_2 sensor devices, a router, and Personal Area Network (PAN) co-ordinators. All the devices are based on MCU and ZigBee chip. The sensor devices measure SPO_2 data from patients and transmit the data to the router. The router sets schedule for data transmission to each device by using a hierarchical routing. The proposed system also contains a web-based management system so that the patient data can be published in the web.

A real-time rehabilitation platform for patients and aged people has been proposed in [20]. The proposed system can collect data about the real-time walking acceleration of the patients. By analysing the gait sequences the computer based rehabilitation system can figure out the normal gait and abrupt falling of the people. The system also includes an ECG detector to monitor the health condition of the patients and the aged people.

A prototype of smart sniffing shoes has been designed for monitoring the foot health of a patient in [21]. The proposed system consists of chemical sensor array installed inside the shoe. The ZigBee technology has been used for the data communication. A technique called principal component analysis (PCA) has been used to monitor the foot health of a patient.

Another ZigBee based health monitoring system has been proposed to monitor temperature, heart rate, blood pressure, and movements of patients in [22]. The main component of the system is an electronic device worn on the wrist or finger of a high risk patient. The system uses a number of sensors including an impact sensor to detect the fall of a person. The system can monitor a medically distressed person and send an alarm to a caretaker system connected to a remote computer.

A similar work for monitoring the patient's pulse has been presented in [23]. The proposed system can monitor the pulses of a patient from a remote location and it can also administer necessary medical treatment. The proposed system consists of a pulse sensor, ZigBee module, and ATmega218P microcontroller. The pulse measured by the sensor is sent to a coordinator through a ZigBee interface. The test results show that the proposed system can cover up to 30 meter distance.

A prototype model for cardiovascular activity and fitness monitoring system based on IEEE 11073 family has been proposed in [24]. The IEEE 11073-10441 defines the set of protocols for tele-health environment at application layer and the rest of the communication is taken care of by the medical grade ZigBee network. The test results show that the proposed system can report severe cardiovascular malfunctioning without compromising the mobility.

A remote heart sound and lung sound monitoring system has been proposed in [25]. The authors have solved the problem related to simultaneous transmission of heart and lung sounds. Sensors have been used to collect the heart and lung sounds and then FastICA is used to separate these two signals. The sound signals are then sent to a remote location via internet for diagnosis.

A low cost sleep monitoring system based on polysomnography has been proposed in [26]. The authors have introduced some innovative sensor pillow and bed sheet system that employ the ZigBee wireless network. To monitor the respiration of the patient a sensor array of force sensitive resistors (FSR) based on polymer thick film device has been used. The sensor array is able to classify and verify the respiratory rate during sleep.

A portable ECG monitoring system has been proposed in [27]. The proposed system is controlled by MSP430 single chip computer, which amplifies and filters the patient's ECG signals and sends data to a central controller using a ZigBee wireless transmission module. Another similar prototype of a ZigBee based ECG signal monitoring system has been proposed in [28]. A PC based GUI interface has been developed to provide ECG signal processing task and health care video tracking and management functions.

In this work we have proposed a remote health monitoring system based on the ZigBee technology and LabView software. The proposed system can monitor ECG signals, muscle power, temperature, and heartbeat of a patient from a remote location. In contrast to other related works we used National Instruments' LabView software for implementing the project. The LabView constitutes a graphical programming environment that can acquire data (i.e., biomedical signals). The LabView relies on graphical symbols rather than textual language to describe programming actions. The principle of dataflow governs program execution in a straightforward manner. We chose LabView software because it is easy to program and it has powerful data acquisition system. In addition the output data generated by the LabView program can be easily acquired into hardware. The data acquisition is performed by using Data Acquisition System (DAQ) provided by the National Instrument. The use of DAQ reduces the complexities of the circuits. Since the major portion of this work is implemented in LabView, the proposed system can be easily reconfigured and adapted to accommodate more options in future. The system can send data to a remote location for diagnosis. The system can also publish data in the internet so that the concerned healthcare professionals can monitor their patients from anywhere around the World at any time.

3. SYSTEM MODEL

The proposed system consists of a set of biomedical sensors attached with the body of a patient. A wireless transmitter is used to send the data to a wireless receiver connected to a local monitoring unit. In this work we used six biomedical sensors to monitor heart beat rate, temperature, changes in muscles power, and ECG signals of a patient. These sensors convert the physiological changes of the patient's body into biomedical signals. The conditioning circuit (i.e., Arduino microcontroller) reads the data from the sensors and controls the transmission of data to a monitoring unit. The monitoring unit displays the data that is used by the physicians for necessary medical advices. The wireless receiver consists of Xbee that receives data and sends it to the local monitoring unit. The monitoring unit can display, record, and analyze the data. It can send reports as well as alarming messages to the healthcare professionals. The system block diagram of the proposed system is shown in Figure 2.

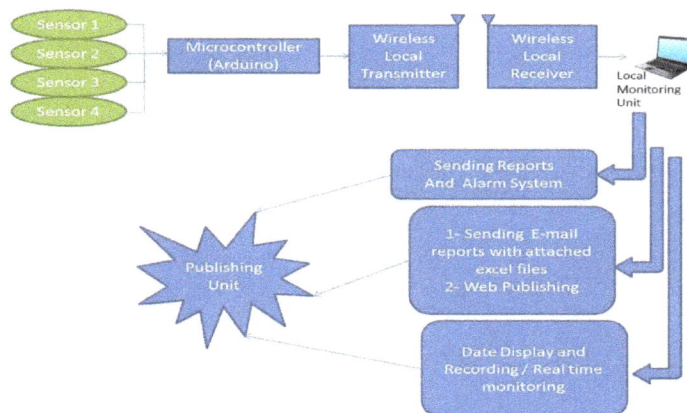

Figure 2 The system block diagram.

Based on the customer requirements the system hardware can be easily modified to accommodate more sensors. The data transmission can also take place via wired or wireless channels. The proposed system can be connected to the Internet for global communication. In addition, the proposed system is carefully implemented in hardware and software system so that it can be adapted to fulfill the user's requirements. Since accuracy is one of the most important issues in biomedical signal processing, the proposed system has been field tested extensively to ensure its accuracy. The system can continuously monitor the health of a patient twenty four hours a day. The proposed system is also able to inform the healthcare professionals about any unusual health conditions of a patient. The doctors can also use the publishing system incorporated with the system. When the measured data exceeds the allowable normal range, the system can send an alarm message to the concerned healthcare professionals. The system can facilitate healthcare professionals to perform immediate medical diagnosis and to administer the medical treatment if needed. The system measures different physical parameters of a patient by using four different sensors as shown in Figure 3. The microcontroller receives the signals from the sensors and processes them before sending them to a ZigBee transmitter module. The transmitter module transmits the signal that is received by the receiving antenna of the ZigBee receiver.

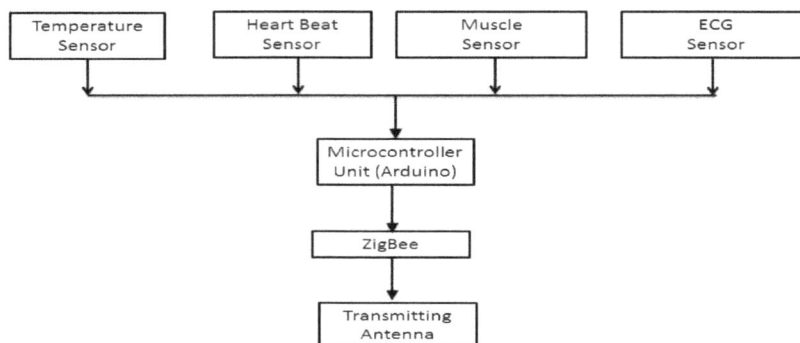

Figure 3 System block diagram of transmitter section

The system block diagram for the receiver is shown in Figure 4. The receiver antenna receives the data sent by the transmitting antenna and then the data are sent to a PC (i.e., Monitoring Unit) for display. The Monitoring Unit sends report using the internet to the concerned healthcare professionals.

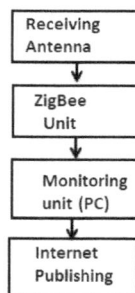

Figure 4 System block diagram receiver section

The temperature sensor used in our system is shown in Figure 5(a) and the associated program flowchart in shown in Figure 5(b). We used LM35 sensor for our project. The LM35 is a high

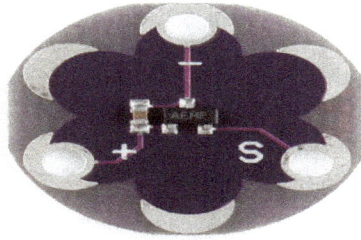

(a) Temperature Sensor

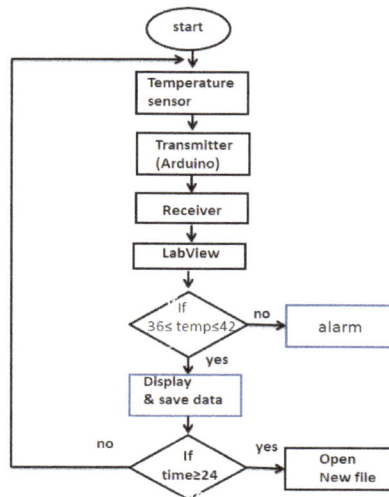

(b) Flowchart

Figure 5 The temperature measurement.

precision integrated temperature sensor. It generates an analog voltage depending on the temperature of the patient's body. The sensor output voltage is linearly proportional to the body temperature. The sensor circuitry is sealed and is not subject to oxidation. The LM35 generates a higher output voltage than thermocouples. The sensor can measure temperature and generate signal that is sent to a microcontroller. The data are then transmitted by the ZigBee to the PC. The sensors are connected to the I/O port of the PIC microcontroller (i.e., Arduino). The output voltage is converted into temperature by a simple conversion factor. As shown in Figure 6 the temperature sensor measures the temperature and converts it into electrical signal. The electrical signal is then processed by a microcontroller and the LabView software. Finally, it is displayed in the monitoring unit. We set the normal body temperature of a patient in the range of 36°C - 40°C. If the temperature reading is less than 36°C or more than 40°C degree the alarm will be ON and it will send an alert message to the concerned healthcare professional.

The heart beat sensor used to measure the heartbeat of the patient is shown in Figure 6(a). This sensor monitors the flow of blood through a clip that is attached with a fingertip. The sensor has a laser that emits light through the skin and measures the reflection of the laser due to the flow of the blood. The heart beat rate of an individual may vary. At rest, an adult man has an average pulse rate of 72 beats per minute. Athletes normally have a lower pulse rate compared to that of a less active people. On the other hand children have a higher pulse rate (approx. 90 beats per minute). We set the critical pulse rate at 120 beats per minute. The flowchart for the heart pulse sensor is shown in Figure 6(b). The sensor measures the heart beats and converts them into electrical signals. Then the receiver antenna sends the data to the Monitoring Unit for displaying the data. If the pulse rate is less than 120 beats per minute, the alarm will be ON and it will alert the concerned authority.

(a) Heart beat sensor

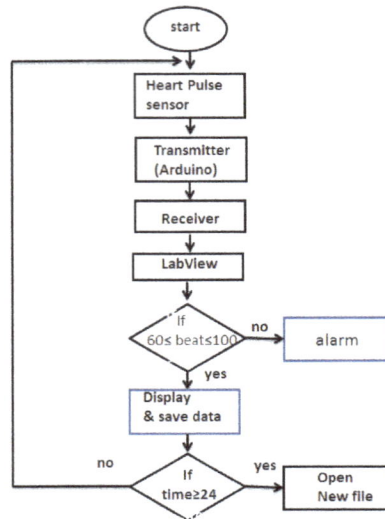

(b) Flowchart

Figure 6 The measurement of heart beat rates

The muscles sensor used in this project is shown in Figure 7(a). The sensor detects the changes in the muscle force and converts it into variable resistive readable values. The local monitoring unit coverts the resistive date from the muscle sensor to a power signal. The main purpose of the muscle sensor is to measure the power from the muscles. In some cases elderly people who are unable to move or disabled and their muscles power decreases over a time period, the sensor sends an alert message to health care service provider. If the patient is moving or standing still or sleeping, the status of the patient's movement can also be monitored by our proposed system. The flowchart of the muscle sensor is shown in Figure 7(b). The transmitter sends the signal which is received by the receiver. The receiver sends the data to the Monitoring Unit for graphical display. If the reading is less than 150 or more than 500 the alarm will be ON and it will alert the health care service provider.

(a) Muscle sensors

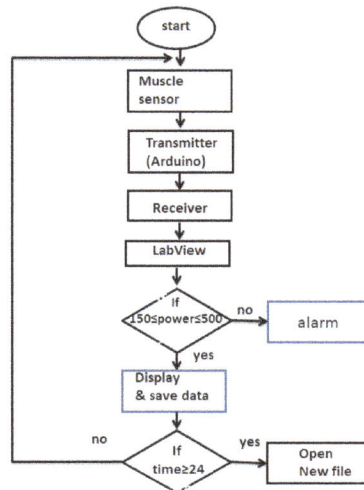

(b) Flowchart

Figure 7 The measurement of muscle power

ECG is an important biomedical parameter and is used clinically in diagnosing various diseases and conditions of a patient's heart. The acquisition of a real time ECG signal requires an expensive CARDIART machine and only experienced cardiologist can interpret the ECG signal. However, in developing and under developed countries people cannot make use of this facility because they live in remote areas. In this proposed system we developed an accurate, low cost, and user friendly real time ECG acquisition system for monitoring general cardiac abnormalities. By using the proposed system the ECG data can be monitored and analysed from a remote location via Web browser. Hence, the system supports long distance diagnosis. The system can generate a report of the patient's condition by using the ECG data. The hardware ECG is mainly consisting of an electronic circuit, which uses amplifiers and switches to amplify the readings from the sensor so that it can be read and displayed. The software ECG monitors the readings from the sensors and then converts them into the readings based on a defined formula. The reading of the software ECG has been verified with that of a CARDIART machine in order to ensure reliability and accuracy. The proposed system measures ECG data from four sensors placed at four different places of the patient's body as shown in Figure 8 and provides an output as shown in Figure 9. The detail operation of the ECG is illustrated in Figure 10. First, the microcontroller (i.e., Arduino) takes the signals from the sensors and converts them into readable values using some defined formula. Then the data are transferred from the transmitter to the receiver (i.e.,Xbee). Then the LabView program combines these readings from the four sensors and provides one variable output. The ECG signal flow diagram is shown in Figure 10.

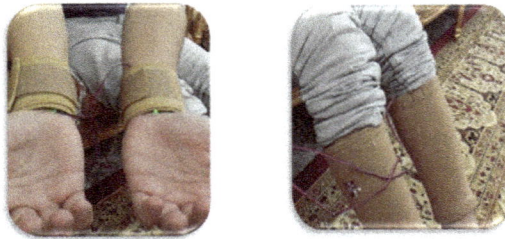

Figure 8 ECG Sensors

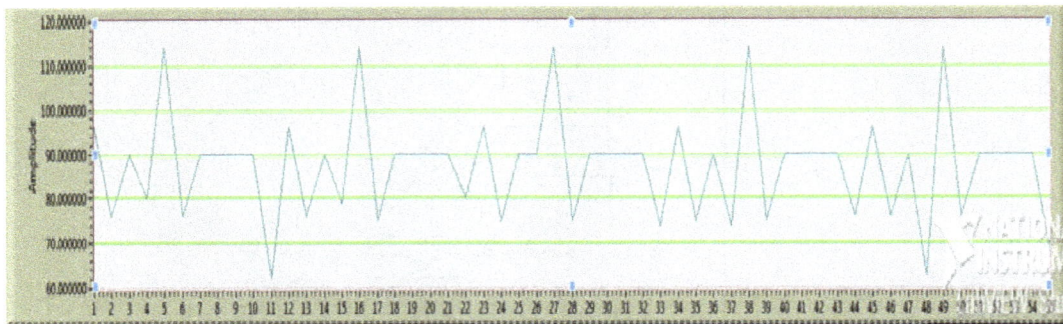

Figure 9 The output of ECG sensors.

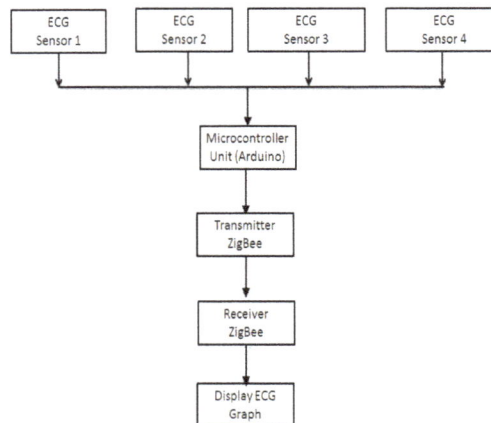

Figure 10 ECG signal flow diagram

4. IMPLEMENTATION AND RESULTS

The proposed system provides a patient with continuous health monitoring service. The signals generated by the sensors are processed by a built-in microcontroller. The processed data are then transmitted by ZigBee wireless transmitter. Finally the received data is sent to a PC. The proposed system works as follows: (a) the user will wear the sensors, (b) the sensors will start

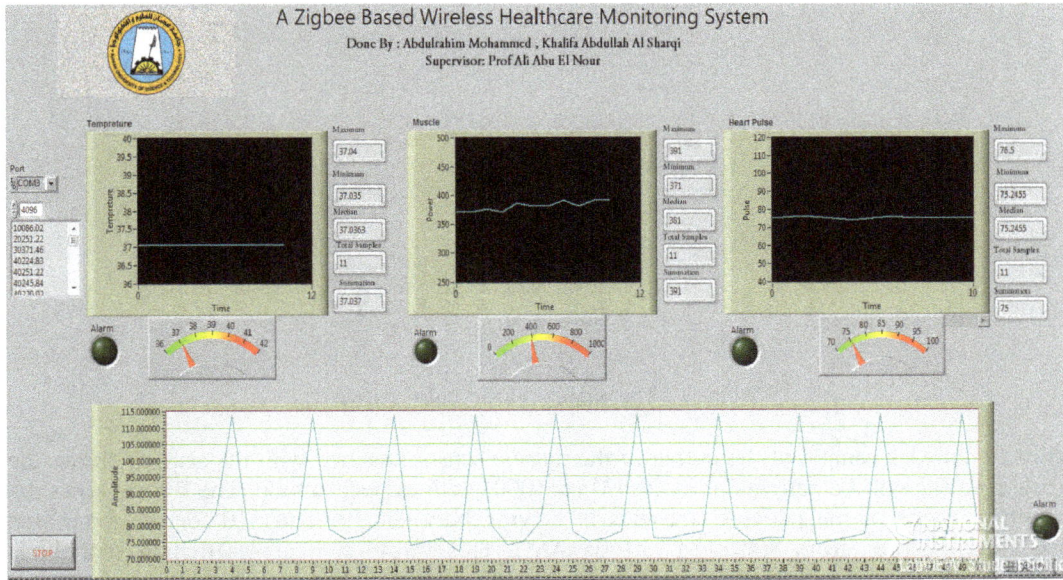

Figure 11 Labview front panel

reading the temperature, heart beats rate, muscle power, and ECG data, and (c) the program will send the monitored data wirelessly and interfaces between the LabView software and the hardware. The front panel of the proposed system is shown in Figure 11.

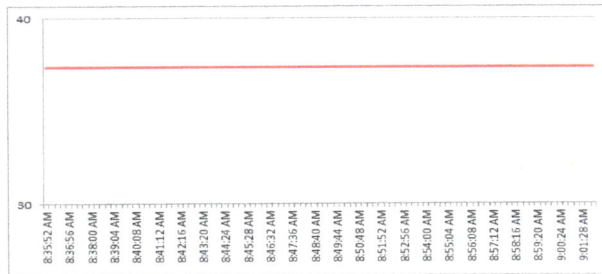

Figure 12 Output of the temperature sensor

Figure 12 shows a sample of the temperature sensor data. It shows the variation of the temperature with respect to time. It is depicted in this figure that the temperature is in the normal range from 8:35 AM to 9:01 AM which means that the patient's body temperature is within the normal range. Since the body temperature is within the set normal range, the system will not send any alarm message to the healthcare professional.

Figure 13 shows a sample output of the heart pulse sensor. The output shows that there are the changes in the heart pulse rate from time to time. The figure illustrates that the patient heart pulse rate varies between 75 beats per minute and 76 beats per minute for the monitored period of time. Since the heart beat rate is within the normal limit, we assume that the patient is at stand still condition.

International Journal of Wireless & Mobile Networks (IJWMN) Vol. 6, No. 3, June 2014

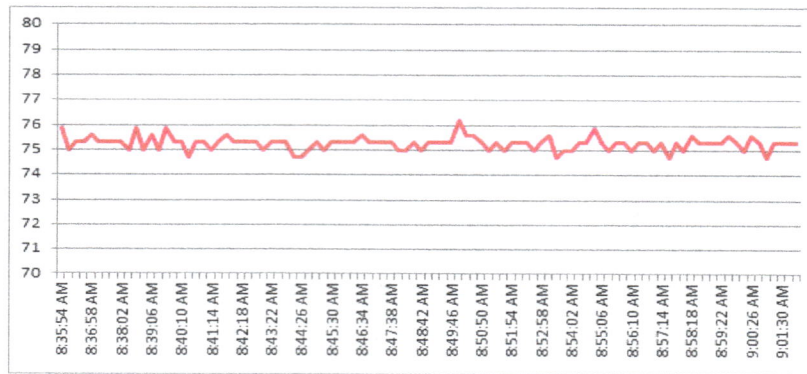

Figure 13 The output of heart pulse sensor

Figure 14 shows a sample of the output of the muscle sensor attached with the arm. It shows the percentage usage of the patients' muscles (Max=1000) with respect to time. The figure shows the percentage usage of the patients' muscles is varying between 200 and 310 for the monitored period of time (i.e., from 8:40 AM to 9:01 AM). This variation indicates that the patient is moving and hence the muscles power is increasing with respect to time.

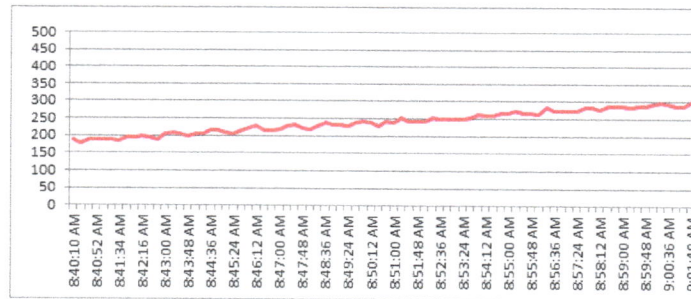

Figure 14 The output of the muscle sensor (Arm)

5. CONCLUSIONS

A reliable wireless healthcare monitoring system has been designed and successfully implemented in this work. The proposed system has been field tested. The test results show that the proposed system is able to monitor the body temperature, heart pulse rate, ECG signal, and muscle power with enough accuracy. Since the proposed system is based on ZigBee, we can conclude that it is a low power and low cost system. Moreover, major part of the proposed system has been implemented in using LabView software. Hence, the proposed system is easily reconfigurable and it can be connected to the Internet easily. The system is also able to store physiological data of patients for 24 hours a day and seven days a week. In future the proposed system can be extended to include more sensors that can measure more parameters like diabetes and blood pressure. The proposed system is flexible enough to include such kind of modifications.

REFERENCES

[1]   Linking Population, Poverty, and Development available at www.unfpa.org.

[2]   Health United States, 2004 available at http://www.cdc.gov.

[3]   Projected Population of the United States, by age and sex:2000 to 2050 available at http://www.census.gov

[4]   Medical Costs available at kff.org/health-**costs**

[5]   Home Digital Home: Wi-Fi Gets a Place At The Table available at http://www.wirelessfidelitymag.com/

[6]   Bonam Kim, Youngjon Kim, InSung Lee, and Ilsum You, " Design and Implementation of a Ubiquitous ECG Monitoring System Using SIP and the ZigBee Networks", In the proceedings of the Future Generation Communication and Networking (FGCN 2007), December 6-8, 2007, Jeju, Korea, pp. 599-604

[7]   Fariborz H., Moghawemi, M., and Mehrkanoon, S, " The design of an intelligent wireless sensor network for ubiquitous healthcare", In the Proceedings of the International Conference on Intelligent and Advanced System, November 25-28, 2007, Kualalumpur, Malaysia, pp. 414-417

[8]   Xi Xueliang, Tao Cheng, and Fang Xingyuan, " A health care System based on PLC and ZiGbee", In proceedings of the International Conference on Wireless Communication, Networking and Mobile Computing, September 21-25, 2007, Shanghai, China, pp. 3063-3066

[9]   Manohar, A. and Bhatia, D., " Pressure Detection and Wireless Interfaces for patient bed, " In the Proceedings of the IEEE Biomedical Circuits and Systems Conference, November 20-22, , 2007, Baltimore, MD, pp. 389-392

[10]   Xiao Hu, Jianping Wang, Qun Yu, and Waixi Liu, " A Wireless Sensor Network Based on ZiGbee for Telemedicine Monitoring System", In the Proceedings of the 2nd International Conference on Bioinformatics and Biomedical Engineering, May 16-18, 2008, Shanghai, China, pp. 1367-1370.

[11]   Juang J.Y., and Lee, J.W., " ZigBee Device Access Control and Reliable Data Transmission in ZigBee Based Health Monitoring System", In the Proceedings of the 10th International Conference on Advanced Communication Technology, February , Guagwan-Do, 2008, pp. 795-797

[12]   Hsu Chih-Jen, " Telemedicine information monitoring system", In the Proceedings of the 10th International Conference on E-health Networking, Application, and Service, Julu 7-9, 2008, Singapore, pp. 48-50.

[13]   Jingram Luo, Yulu Chen, Kai Tang, Juwen Luo, " Remote monitoring information system and its applications based on the Internet Things", In the proceedings of the International Conference on Future Biomedical Information Engineering, December 13-14, 2009, Sanya, pp. 482-485

[14]   Ramos, J., Austin, J.L., Torelli, G., Duque-Carnillo, J.F.," A wireless sensor network fort fat and hydration monitoring by bioimpedance analysis," In the Proceedings of the 6th International Wearble Micro and Nano Technologies for Personalized Helath, June 25-26 , 2009, Oslo, pp. 49-52

[15]   Yuanlong Liu, and Sahandi, R.," ZigBee network for remote patient monitoring on general hospital wards", In the Proceedings of the XXII International Symposium on Information, Communication, and Automation technologies, October 29-31, 2009, Bosnia, pp. 1-7

[16]   Chengcheng Ding, Xiaopei Wu, Zhao Lu, " Design and Implementation of the ZigBee-based Body Sensor Network System", In the Proceedings of the 5th International Wireless Communciations, Networking, and Mobile Computing, September 24-26, 2009, Beijing, pp. 1-4.

[17]   Martin, H., Bernades, A.M., Bergesio, L., and Tarrio, P., " Analysis of key aspects to manage wireless sensor networks in ambient assisted living Environments", In the proceedings of the 2nd International Symposium on Applied Sciences in Biomedical and Communciation Technologies, November 24-27, 2009, Brastislava, pp.1-8

[18]   Becher, K., Fugueredo, C.P., Muhle, C., and Ruff, R., " Design and realization of a wireless sensor gateway for health monitoring", In the proceedings of the Annual IEEE Conference on Engineering in Medicine and Biology Society, August 31-Sept 4, 2010, Buenes Aires, pp. 374-377

[19]   Zhou Yan, and Zhu Jiaxing, " Design and implementation of ZiBee based wireless sensor network for remote SPO_2 monitor", In the proceedings of the International Conference on Future Computer and Communication, May 21-24, 2010 Wuhan, pp. V2 278-V2 281

[20]   Tian Lan, and Xiaogiong Li, " Gait Analysis via a high resolution triaxial accelartion sensor based on ZigBee Technology", In the proceedings of the International Conference on Complex Medical Technology, May 25-28, Beijing, 2013, pp. 697-702

[21]   Seesaard T., Lorwongtragol, P., and Nilpanapan, T, " An smart sniffing shoes based on embroided sensor array", In the proceedings of the 10th International Conference on Electrical Engineering/Electronics, Computer, Telecommunication, and Information Tehcnology, May 15-17, 2007, Krabi, pp. 1-4

[22]   Deppa, A., and Kumar P.N., " Patient health monitoring based on ZigBee Module", In the proceedings of the International Conference on Optical Imaging Sensor and Secuirty, July 2-3, 2013, Coimbatore, pp. 1-4

[23]   Niswar, M., Ilham, A.A., Palantei, E, and Sadjad, R.S., " Performance evaluation of a ZigBee based Wireless sensor network for monitoring patient's pulse status", In The proceedings of the International Conference on Information technology and Electrical Engineering, Yogyakarta, Octiber 7-8, 2012

[24]   Gangwar, D.S., "Biomedical sensor network for cardiovascular fitness and activity monitoring" In the proceedings of the IEEE Point of Care Healthcare Technolonogies (PAT), January 16-18, 2013, Bangalore, pp. 279-282

[25]   Yi Zhang, Sixuan Chen, Jin Wu, Yuan Luo, " Remote heart and lung sound monitoring system design based on ZigBee", In the proceedings of the 6th International Conference on Biomedical Engineering and Informatics, October 16-18, 2012 Chongqing, pp. 1109-1111. s

[26]   Lokavee, S., Puntheeranurak, T., Kerdeharoen, T., Wahanwisuth , W.," Sensor pillow and bed sheet system: Unconstrained monitoring of respiration rate and posture movements during sleep", In the Proceedings of the IEEE Conference on Systems, Man, and Cybernetics, October 14-17, Seoul, 2012, pp.

[27]   Hongli Yang and Jhing Chai, " A portable wireless ECG monitoring system based on MSPCBOFGA39" In the proceedings of the International Conference on Intelligent Computation and Biomedical Instrumentation, December 14-17, 2011, Wuhan, pp. 148-151

[28]   Hui-Yang H Sia, Linag-Hung Wang, Feng-chi Lin, and Chien-Chou Chen, " Design and implementation of a wireless ECG acquisition and Communication system with health care services", In the proceedings of the International Symposium on Bioelectronics and Bioinformatics, November 3-5, Suzhou, 2011, pp. 25-28

Permissions

All chapters in this book were first published in IJWMN, by AIRCC Publishing Corporation; hereby published with permission under the Creative Commons Attribution License or equivalent. Every chapter published in this book has been scrutinized by our experts. Their significance has been extensively debated. The topics covered herein carry significant findings which will fuel the growth of the discipline. They may even be implemented as practical applications or may be referred to as a beginning point for another development.

The contributors of this book come from diverse backgrounds, making this book a truly international effort. This book will bring forth new frontiers with its revolutionizing research information and detailed analysis of the nascent developments around the world.

We would like to thank all the contributing authors for lending their expertise to make the book truly unique. They have played a crucial role in the development of this book. Without their invaluable contributions this book wouldn't have been possible. They have made vital efforts to compile up to date information on the varied aspects of this subject to make this book a valuable addition to the collection of many professionals and students.

This book was conceptualized with the vision of imparting up-to-date information and advanced data in this field. To ensure the same, a matchless editorial board was set up. Every individual on the board went through rigorous rounds of assessment to prove their worth. After which they invested a large part of their time researching and compiling the most relevant data for our readers.

The editorial board has been involved in producing this book since its inception. They have spent rigorous hours researching and exploring the diverse topics which have resulted in the successful publishing of this book. They have passed on their knowledge of decades through this book. To expedite this challenging task, the publisher supported the team at every step. A small team of assistant editors was also appointed to further simplify the editing procedure and attain best results for the readers.

Apart from the editorial board, the designing team has also invested a significant amount of their time in understanding the subject and creating the most relevant covers. They scrutinized every image to scout for the most suitable representation of the subject and create an appropriate cover for the book.

The publishing team has been an ardent support to the editorial, designing and production team. Their endless efforts to recruit the best for this project, has resulted in the accomplishment of this book. They are a veteran in the field of academics and their pool of knowledge is as vast as their experience in printing. Their expertise and guidance has proved useful at every step. Their uncompromising quality standards have made this book an exceptional effort. Their encouragement from time to time has been an inspiration for everyone.

The publisher and the editorial board hope that this book will prove to be a valuable piece of knowledge for researchers, students, practitioners and scholars across the globe.

List of Contributors

Nidhi
School of Computer and Systems Sciences, Jawaharlal Nehru University, New Delhi, India

D.K. Lobiyal
School of Computer and Systems Sciences, Jawaharlal Nehru University, New Delhi, India

Bera Rabindranath
Sikkim Manipal Institute of Technology, Sikkim Manipal University, Majitar, Rangpo, East Sikkim, 737132

Sarkar Subir Kumar
Jadavpur University, Kolkata 700 032

Sharma Bikash
Sikkim Manipal Institute of Technology, Sikkim Manipal University, Majitar, Rangpo, East Sikkim, 737132

Sur Samarendra Nath
Sikkim Manipal Institute of Technology, Sikkim Manipal University, Majitar, Rangpo, East Sikkim, 737132

Bhaskar Debasish
Sikkim Manipal Institute of Technology, Sikkim Manipal University, Majitar, Rangpo, East Sikkim, 737132

Bera Soumyasree
Sikkim Manipal Institute of Technology, Sikkim Manipal University, Majitar, Rangpo, East Sikkim, 737132

Barbaros Preveze
Department of Electronics and Communication Engineering, Çankaya University, Ankara, Turkey

Shailender Gupta
Department of Electronics Engineering, YMCA University, Faridabad, India

C. K. Nagpal
Department of Computer Engineering, YMCA University, Faridabad, India

Charu Singla
Department of Electronics Engineering, NGF College of Engg & Technology, Palwal

V. Ramya
Assistant Professor Department of Computer Science and Engineering, Annamalai University, Chidambaram, Tamilnadu

B. Palaniappan
Dean, FEAT, Head Department of Computer Science and Engineering, Annamalai University, Chidambaram, Tamilnadu

Sheeba Armoogum
Department of Computer Science and Engineering, University of Mauritius, Reduit, Mauritius

Vinaye Armoogum
Department of Industrial Systems Engineering, University of Technology Mauritius, Port- Louis, Mauritius

Jayprakash Gopaul
Mauritius Broadcasting Corporation, Reduit, Mauritius

Yashpal Singh
Research Scholar, Mewar University, Rajasthan, India

Kamal Deep
Assistant Professor, Department of Electronics and Communication, JIET Jind, Haryana,India

S. Niranjan
Professor, Department of Computer Science & Engineering ,PDM College of Engineering Bahadurgarh, Jhajjar, Haryana, India

Prasun Chowdhury
Department of Electronics and Telecommunication Engineering, Jadavpur University, Kolkata 700032, India

Anindita Kundu
Department of Electronics and Telecommunication Engineering, Jadavpur University, Kolkata 700032, India

Iti Saha Misra
Department of Electronics and Telecommunication Engineering, Jadavpur University, Kolkata 700032, India

Salil K Sanyal
Department of Electronics and Telecommunication Engineering, Jadavpur University, Kolkata 700032, India

Mohamed A. AboulHassan
Pharos University, Faculty of Engineering, Electrical Eng. Dept., Alexandria, Egypt

Essam A. Sourour
Alexandria University, Department of Electrical Eng., Alexandria, 21544 ,Egypt

Shawki Shaaban
Alexandria University, Department of Electrical Eng., Alexandria, 21544 ,Egypt

S. Alagu
Research Scholar, Department of Computer Science and Engineering Alagappa University, Karaikudi, Tamilnadu, India

T. Meyyappan
Professor, Department of Computer Science and Engineering Alagappa University, Karaikudi, TamilNadu, India

Lokesh. B. Bhajantri
Department of Information Science and Engineering, Basaveshwar Engineering College, Bagalkot, Karnataka, India

N. Nalini
Department of Computer Science and Engineering, Nitte Meenakshi Institute of Technological, Yelahanka, Bangalore. Karnataka, India

S. Gangadharaiah
Department of Information Science and Engineering, Acharya Institute of Technology, Soladevanahalli, Bangalore

Ramprasad Subramanian
Centre for Real-time Information Networks, School of Computing and Communications, Faculty of Engineering and Information Technology, University of Technology Sydney, Sydney, Australia

Shouman Barua
Centre for Real-time Information Networks, School of Computing and Communications, Faculty of Engineering and Information Technology, University of Technology Sydney, Sydney, Australia

Sinh Cong Lam
Centre for Real-time Information Networks, School of Computing and Communications, Faculty of Engineering and Information Technology, University of Technology Sydney, Sydney, Australia

Pantha Ghosal
Centre for Real-time Information Networks, School of Computing and Communications, Faculty of Engineering and Information Technology, University of Technology Sydney, Sydney, Australia

Kumbesan Sandrasegaran
Centre for Real-time Information Networks, School of Computing and Communications, Faculty of Engineering and Information Technology, University of Technology Sydney, Sydney, Australia

Mbida Mohamed
Emerging Technologies Laboratory (VETE), Faculty of Sciences and Technology Hassan 1st University, Settat, MOROCCO

Ezzati Abdellah
Emerging Technologies Laboratory (VETE), Faculty of Sciences and Technology Hassan 1st University, Settat, MOROCCO

Thoraya Obaid
Department of Electrical Engineering, Ajman University of Science and Technology, Fujairah, United Arab Emirates

Haliemah Rashed
Department of Electrical Engineering, Ajman University of Science and Technology, Fujairah, United Arab Emirates

Ali Abu El Nour
Department of Electrical Engineering, Ajman University of Science and Technology, Fujairah, United Arab Emirates

Muhammad Rehan
Department of Electrical Engineering, Ajman University of Science and Technology, Fujairah, United Arab Emirates

Mussab Muhammad Saleh
Department of Electrical Engineering, Ajman University of Science and Technology, Fujairah, United Arab Emirates

Mohammed Tarique
Department of Electrical Engineering, Ajman University of Science and Technology, Fujairah, United Arab Emirates

Khalifa AlSharqi
Department of Electrical Engineering, Ajman University of Science and Technology, Fujairah, United Arab Emirates

Abdelrahim Abdelbari
Department of Electrical Engineering, Ajman University of Science and Technology, Fujairah, United Arab Emirates

Ali Abou-Elnour
Department of Electrical Engineering, Ajman University of Science and Technology, Fujairah, United Arab Emirates